Author's Biographical Notes:

Ramin Amirmardfar was born in 19 March 1971 in Tabriz, east Azerbaijan province, northwestern Iran. He completed him primary and secondary studies in the same city and started studies in1989 in plant protection in Tabriz University. He has continued his studies in the field of Agricultural Entomology in the same University. Beside academic studies he was interested to the Evolution of animals/plants and the effect of gravity of Earth on them. From 1990, started to write papers in this field and nine until of them have been published in scientific Journal. In the year 2000 and 2001 he published two books with the titles "The relationship between Earth gravity and animal Evolution" and "The ABC of Evolution". Married in 2003 and have one daughter.

Preface

In our society the important invention, the scientific value of which is obvious, are accepted by the majority soon, and have a lot of materialistic value for their owners, but those who discover the main principles and scientific basis, do not receive any reward and sometimes their life comes to its end, while not only they have seen no reward but also no one has understood their purpose. But instead, when another invention, comes to market, the fame first group disappears, while the fame of the second group increase, and the importance of the scientific principle increases by the increase of its numbers and its number of applications.

The people who have a lot of knowledge and not adopt their knowledge with each other and also do not relate them with each other, although simple-minded people consider them scientists, but in fact they have no difference with simple ignorant people, and these irrelevant knowledge's have no value. When the knowledge is adopted by a regular and unified growth and is harmonious, it can be called integrated knowledge. This kind of knowledge can not be obtained by the accumulation of some information and assembling of large or small and irrelevant rules, rather they should be studied and analyzed with meekness, so the brain can choose and absorb whatever it needs. When the knowledge and information becomes exclusive and complicated the necessity of their unity of them is being sensed more. If a nervy person can not be found to do such a thing, the world of sciences will be impregnable very soon. Now there are so many proficient who, like a bee are not aware of the work that they are doing. They work very eagerly at a corner and also their work is very useful. But the science is not specific to the results of their individual work.

The growth of the science is like the growth of a living creature; some people should take the trouble of them, and should integrate and combine, so that can successfully unify them. If there is no attempt to collect and unify the knowledge's, the separated scientific facts and small theories will increase, but the science will disappear per se. The person who takes the trouble to understand the very complicated parts of science and to culminate, is like a passerby who watches the desert and whatever is in it, from a mountaintop, the hills with their strange shapes and the thick forests will no more bewilder him, and he can see all of them from the top of the mountain and none of them will prevent the other ones from being seen by him, and he can see all of them and can differentiate and recognize their relationships. It is not necessary for integrated to be broader than other specific sciences, Because the person who wants to acquire it, do not try to be aware of the punctilios and the mysteries which are allocated to exclusive knowledge or don't want to fill his brain with them.

Most of the punctilios which are acquired by the expert scientist a hard work, is of no importance for a comprehensive scientist. As the drawing of the watercourse of a river is very simple for a drawer, to discover which a lot of people have worked very hard. Also for a comprehensive scientist, the record of scientific facts and thoughts is very simple; every one of which is the result of the hard work and intelligence of scientists. But most of scientists prefer not go beyond the experiments and experience, But whatever they become more doubtful, and very soon the most expedient attempts to acquire experimental knowledge, looks like an accidental event for them.

When we compare the determined decision and attempt of a comprehensive scientist with other people's hesitation, we may consider him as a hero. Frankly, in this affair there is some heroic aspect, because it has an aspect of adventure. Exclusive researches usually do not face with failure, because their results are immediate and bring relief. An astronomer who extracts correct calendar for us, and a chemist who provides the colors for us as well as a baker who brings the bread of the oven, is aware of the result of his work, and also to write the cards regularly and putting them in their places carefully or to set the row of the insects and seashells and writing notes and articles about them is satisfactory for most of people. They know that their work will remain forever, because they provide the materials, which are the basis of any scientific compound, by the passage of time the buildings are built with these materials, and perhaps the building will be destroyed but the materials remain! Most of scientists stop here, that are they providing the material but do not build the building. I suppose that they afraid of bewildering and don't go forward following their natural instinct. They have right not to go forward, but the thought that every where they become bewildered, all of them will be involved in this situation, is wrong. If it said that the theories of a comprehensive scientist precede his experiences, the answer is that, this drawback is true about all other scientific theories and the person who poses a theory, should not be questioned that if he himself has tested it or not.

"George Sarton"

Introduction

Elephants, rhinos, kangaroos, camels, all mammals evolved at the first time by small bulk animals. Mammals have managed to overcome gravity by the aid of 4-chambered hearts and strong blood circulatory systems and making their bulks larger. But during the time a gradual increasing of gravity occurred, mammals failed against greater gravity and had to minimize their bulks. The process of minimizing the body size has continued up to now and will continue in the future Mammals will become gradually smaller.

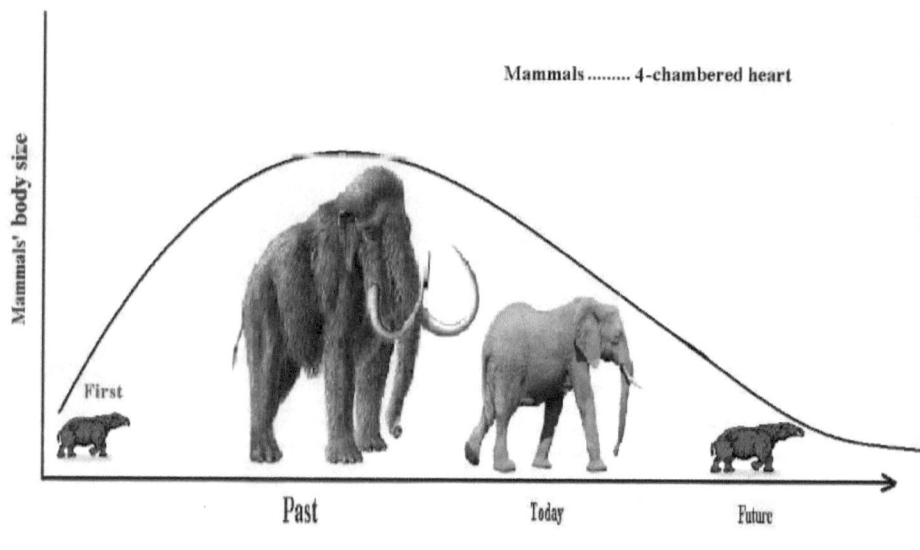

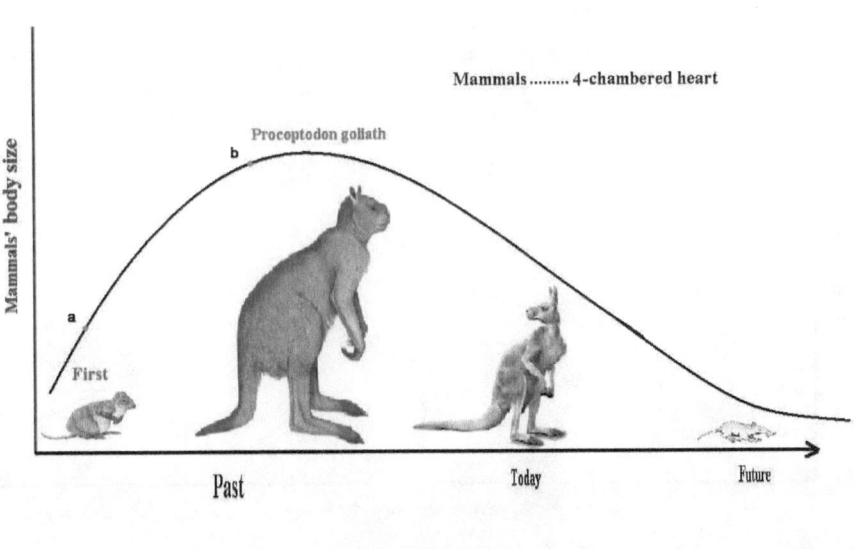

Mammals .... 4-chambered heart

Procoptodon goliath

b

a

First

Past

Today

Future

Earth's gravity is increasing

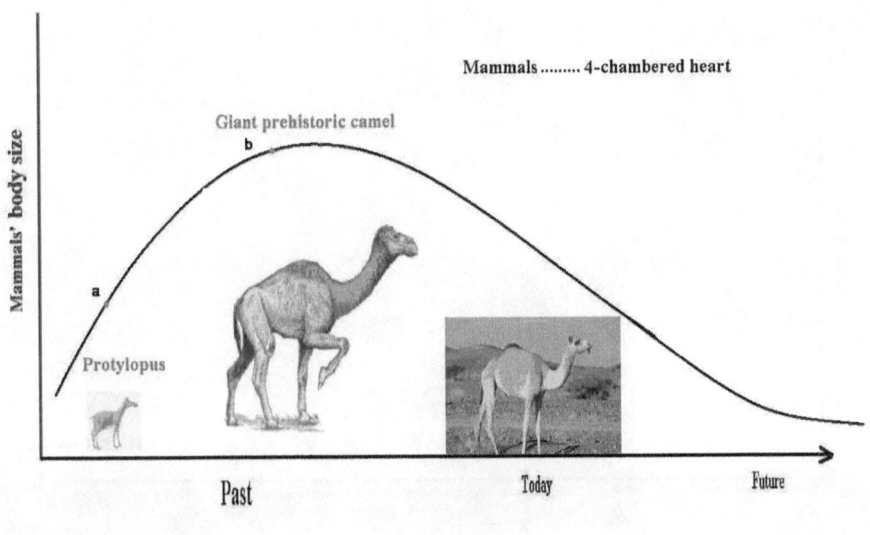

Mammals .... 4-chambered heart

Giant prehistoric camel

b

a

Protylopus

Past

Today

Future

Earth's gravity is increasing

Ramin Amirmardfar

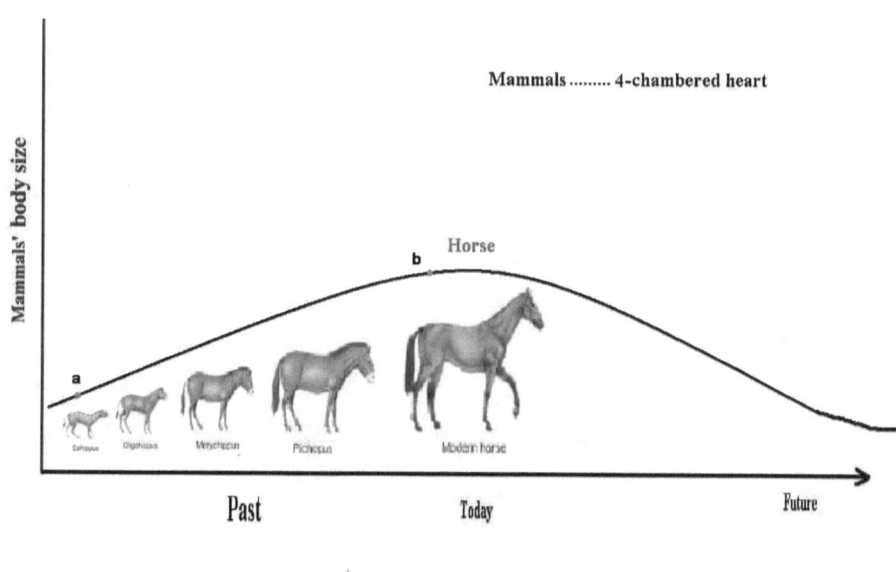

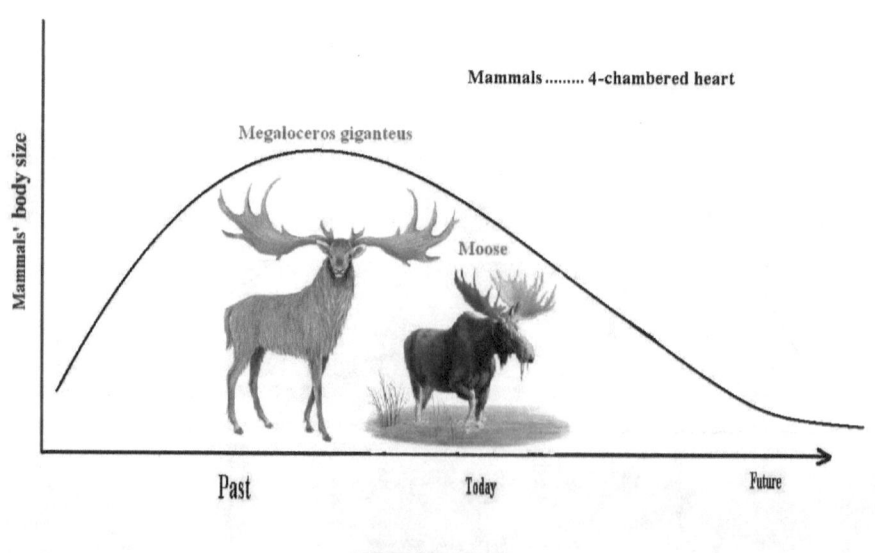

Ramin Amirmardfar

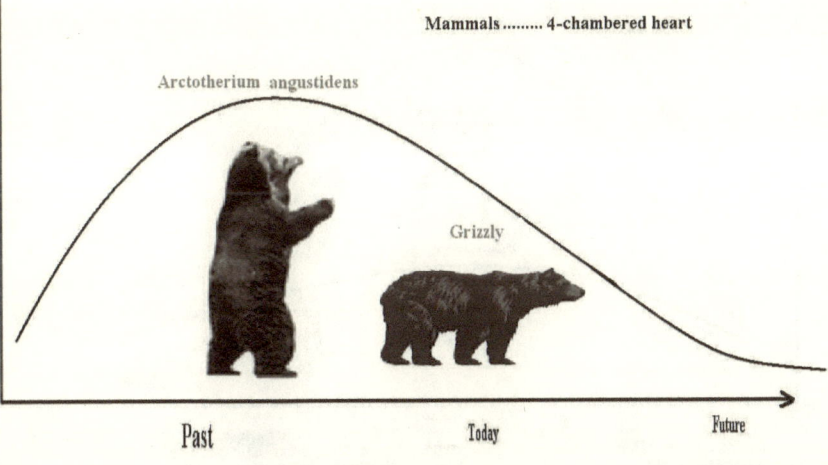

Mammals .... 4-chambered heart

Arctotherium angustidens

Grizzly

Mammals' body size

Past   Today   Future

>------- Earth's gravity is increasing ------->

Ramin Amirmardfar

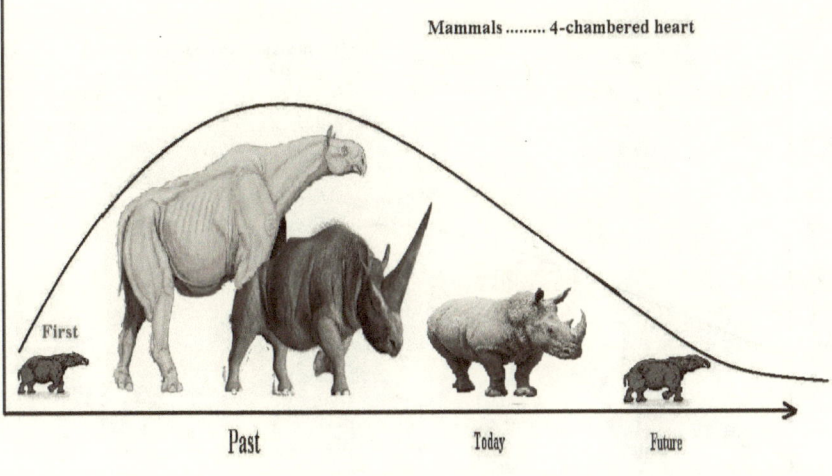

Mammals .... 4-chambered heart

Mammals' body size

First

Past   Today   Future

>------- Earth's gravity is increasing ------->

Ramin Amirmardfar
2017

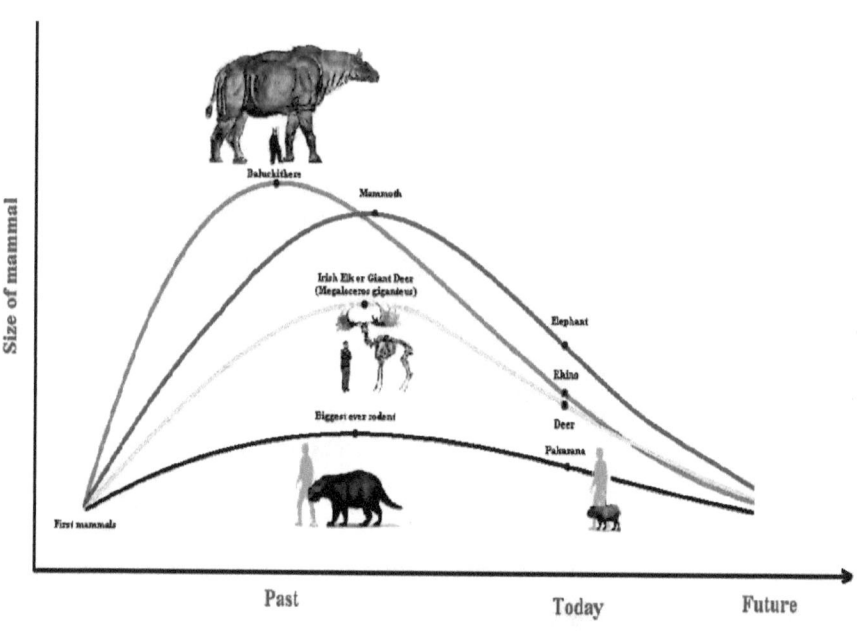

Ramin Amirmardfar

All mammals after its peak again become smaller. But as you can see the peak point have been located them at different times. The peak of giraffe has been located in present tense. This means that giraffes will becomes smaller in the future.

Mammals .... 4-chambered heart

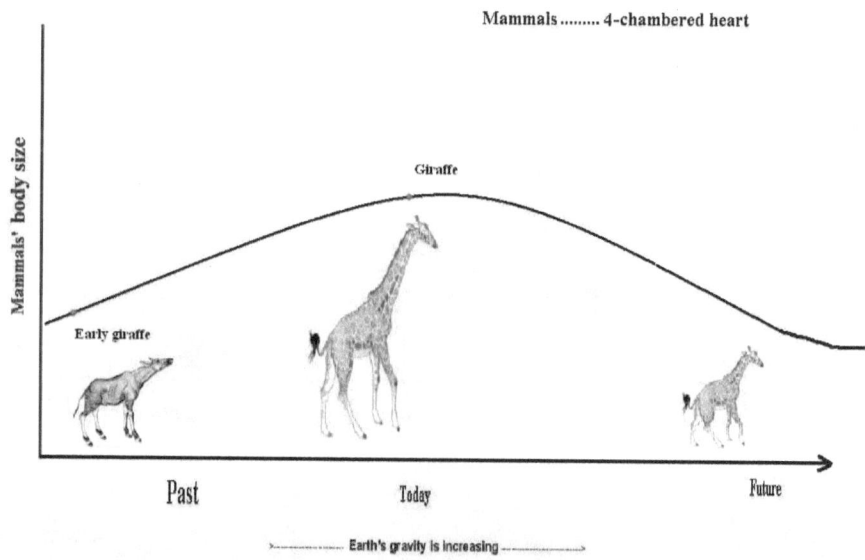

Earth's gravity is increasing

Ramin Amirmardfar

1. Why insects have small bulks?

Before explanation of my new theory, I point out to those theories which have been presented before on this respect.

Exoskeleton:theory: Insects have exoskeleton and this theory point out that this skeleton isn't able to grow more actually, the insect is imprisoned inside this skeleton and isn't able to get larger. Defects of this theory: All of arthropod owns exoskeleton, and some of them like crabs are able to own rather large body which is discordant with this theory. On the contrary, there are some insects that own softer exoskeleton, but again aren't able to enlarge. Also in the past (250 million years ago), there were insects that owned big bodies (like dragonfly by length of 75 cm) and this theory can't explain the reason of their existence. Because they owned hard exoskeleton too but they were able to enlarger!

Respiratory system theory: Insects respiratory system comprise thin and capillary tubes called tracheae that supplies oxygen directly through existent pores on the body, to all insect bodies interior cells. That means, opposite other animals, oxygen isn't been carried by blood. It flows through inside of tracheas by means of propagation phenomenon, and reaches cells. This theory says that the propagation of oxygen inside the cells, is possible only in short intervals and if insects have large body, the length of tracheas will lengthen and oxygen can not reach cells. So insects are forced to own small body. About existence of large insects in the past, this theory predicts the existence of high percentage oxygen in earth's atmosphere. According to this theory, atmosphere's oxygen amount, 250 million years ago, was higher than the oxygen, would make it possible the propagation of it through the long tracheas of insects. Therefore insects at that time were able to be longer than present insects.

My theory is based on animals' blood circulation system. The theory compares the power and evolution of blood circulation system for animals and expresses the relation between bulk largeness and power of blood circulation system. According to this theory as much as blood circulation system is stronger and complete, animals' bulk can be larger (bigger). Insect's blood circulation system is too incomplete and there are no blood vessels. Insect's bodies, with lack of blood vessels, aren't able to supply blood consisting nutrition to remote cells, so their body are forced to be small. In this theory air pressure and gravity, are two physical factors that have impact on function of blood circulation system. The most effort of blood circulation system is used to conquest on gravity. As much blood circulation system is strong, is more able to conquest on gravity and send blood upper into animals' brain. Consequently, animal is able to enlarge more and taller. For example, elephants and giraffes, who own the most evoluted and strongest blood circulation system, are the largest and tallest ones among the animals.

2. Why can not an insect grow to the size of a giraffe?

The creatures of the Earth have different sizes. All of vertebrata, mollusk, worms, insects, belong to the animal kingdom. But we see that their individual kinds show extreme variations in size. The largest animals, such as whales and giraffes belong to the class of Mammalia, and the smallest ones, i.e. insects and Acarina belong to Arthropoda. Largeness and smallness are relative qualities. A thing, on its own, cannot be large or small, and only by comparison with other things, becomes large or small. If we only observed the insects of the world, we couldn't tell that they are small creatures, but when we see a fly resting beside a horse, we become surprised at the great difference of their size, and the question occurs to us: Why are insects smaller than other creatures? "Because of the limited weight of hard exoskeleton, the size of no arthropoda creature encroaches a determined extent." This has been the answer of zoologists and entomologists to the above question. But the answer isn't satisfying even for themselves. Because the body of many insects is soft, but they yet are very small, and against those are crayfishes whose exoskeleton is very hard, yet are greatly larger than insects! Entomologists compare the body structure of an insect with that of a large animal such as a giraffe. They are searching for factor in bodies of insect, which prevent their getting big. But till now they have had little success. For finding an appropriate answer we must look in another direction. In other words, we must instead of searching for the preventive factor of growth of insect, search within the system in the body of the giraffe, which the insects lack, and which contributes to greater size of the body. The head, which carries the brain, is the most important part of any creature's body. The brain is important and sensitive, and must receive enough water and food all the time. Providing necessary water and materials to the brain is among the primary responsibilities of circulatory system. The heart, like a pump, turns the blood, which carries water, and nutritive substances and the vessels carry the blood to the brain, by use of other organs, which act like tubes. Suppose we want to send water from a tank, which is placed on the lower part of a structure to upper floors for usage of resident of the structure. But if we do not use tubes (plumbing) and simply tried to carry on the job only with a pump, it would be a difficult task. The pump can not carry the water to regarded points of the upper floors without tubes and plumbing, and only to a limited extent can the pump shoot it upwards. In this case, the water not only won't reach to desired points of upper floors of the structure, but also at the lower floors, instead of arriving at its desired points (bath and kitchen) it will flood into all the rooms and walls. Without plumbing tubes, we couldn't make high structures functional, and we would be convinced to construct low ones in order to distribute the water for its residents. Within the body of the giraffe, there is not only a strong pump (heart), but also many "plumbing" tubes (arteries, veins, capillaries), so its body can convey blood to the highest points where the brain is located, and to other organs and systems, easily as well. Any organ of the body of a giraffe can receive as much blood as it needs. When the animal eats food, the blood is succinctly guided to its digestive system, operates as the feeding organ. Any cell of the body of the giraffe can receive necessary amounts of water and other substances from nearest capillary systems and send redundant and unnecessary substances to be discarded, through the same system. But the body structure of an insect possesses an entirely different system. There is no "plumbing" to distribute the "water", but only a pump (heart), which carries the blood from the dorsal and pushes it forward. There isn't any upper floor in this structure, and at the lower rooms, the water, instead of flowing in tubes, fills all the rooms, and the residents are basically flooded in water. Individuals in the lower floors of an insect, not only receive food from the "water", but also pour the unnecessary substances into it. The insects have an open circulatory system. This means that there are no vessels or "plumbing" within their body, and the blood moves openly through it. There, every place is full of blood, and organs and systems of the body are "drowning" in it. They receive water and necessary substances from it, and pour the unnecessary excrament into it. The heart, for the lack of vessels, cannot convey the blood to far distances and high points, so all the systems of the body have to gather around the heart, so as not to suffer the shortage of water and necessary substances. The circulatory systems of insects have little authority on the rate of distributing the blood to different organs, and organs which are near the heart, receive more necessary materials and enjoy preferable conditions. The brain, being the most important, is located in front of the aorta, where the blood first emerges from the heart. The heart always takes the blood from the back of the body and pushes it forward towards the head. If the brain of insect was at a high point, like the giraffe's brain, the heart couldn't send blood to it for lack of vessels, so the brain and other organs of insect have to be near the heart. In other words, the size of the insect must be small. In this manner, it is obvious that any creature wanting to grow in size, must have the necessities of distributing the blood, and because the insects lack such means, they cannot grow in size. In other words, the insects, by lacking the "plumbing" vessels, have small sizes, and the giraffes having so many vessels can then become comparatively large. On the whole it can be said that any creature which has a more complete circulatory system, will become larger than others. To help make the matter more clear, it is noteworthy to compare the circulatory system and size of some other creatures also. There are species in the worm groups which, on the whole, lack the circulatory system and are very small and microscopic, i.e. Nematoda and Bryozoa. Some of worms have a simple circulatory system consisting of some linear vessels and a small heart. These species can make themselves a little larger, such as Phoronidea, Sipunculoidea and Brachiopoda. Among the worms, only the segmented ones have an advanced circulatory system, consisting of linear and partial vessels, some hearts and capillaries. Their blood circulation is closed. The segmented worms, having such a circulatory system are the largest species among worms, i. e. Rhinodrilus Fafneri which reaches 210cm long and 2.5cm diameter, and Eunice Gigantea which is 3m, long. In the Mollusca group, we see a variety of circulatory systems, both open and closed. The individuals belonging to open circulatory system, and having few vessels and a simple heart, such as Amphineura and Gastropoda are smaller. But individuals whose circulatory system is more advanced, such as Bivalves, are of a larger size. Finally, Cephalopoda, who are the most developed ones of Mollusca, and have a closed circulatory system with many vessels and heart chambers, are larger than other Molluscas, such as large Octopuses and Architeuthis, which have a body of 2.5m long and arms as longs 12-18m. All members of the Arthropods group have an open circulatory system, so they can not grow very much. The extent of the evolutionary process, however, is not the same among all individuals of this group. Among the arthropods, Crayfishes are the largest species, because their circulatory system, in addition to their hear,t has many arteries and the heart can guide the blood to relatively far distances within them. But because of lacking capillaries and veins, the blood is distributed in its coelum "body cavity". Also, the circulatory system is open. The Circulatory system of Scorpions and some other Arthropods has fewer arteries, so they are smaller than Crayfishes. All insects lack vessels, and only have a Vaisseau Dorsal, which consists of some ventriculites and end in a short aorta. So insects must remain a small relative size. Among insects, those which have longer Vaisseau Dorsal, such as Cockroaches and Grasshopers are larger than ones whose Vaisseau Dorsal is shorter, having less ventriculites, i.e. Coccidaes. The size of the smallest insects is less than the greatest Monocells, i.e. 0.25mm. But these insects are not the smallest Arthropods. Acarina are the smallest ones. Parasite Vegetable Acarias are even smaller that 0.1mm. For observing Acarias of Eryophidae group, magnifying instruments are needed and used. So these small Acarias must have some simpler circulatory system than even insects. Larger Acarias, like insects, have a Vassea Dorsal and heart. But small Acaria lack such organs and their circulation is made only by body muscles and movements of internal organs such as the digestive system. Among Vertebrata, the ones who have a more complete circulatory system, are the larger creatures, i.e. Mammalia, whose circulatory system is the most complex and complete, so they become the larger species such as the whale, elephant and giraffe. So as a whole we conclude that any creature living on earth, establishes a direct relationship between its circulatory system and its relative size. No creature on Earth is an exception to this principle.

3.Relationship between the power of the blood/fluid circulatory system and size of animals. Crocodiles are not a reptile.

I have found this matter for the first time that each animal or plant that has a stronger blood/ fluid circulatory system could have a larger bulk.

Today's insects can never grow to the size of larger animals because they lack the pipes and pumping system necessary to counteract the influence of gravity. Traditionally, the problem of the exoskeleton and moulting are invoked to explain the limited growth of insects, however this point overlooks that early insects grew to much greater size than those of today. Meganeuropsis permiana, an ancient dragonfly, measured an impressive 71 cm from wing tip to wing tip. In my theory embraces the important relationship between structure, gravity and function. The size of an organism is contingent upon the type of its circulatory system. One can think of this as a scale beginning with the smallest animals and ending with the largest. The smallest of animals lack a circulatory system; the slightly larger ones have an open circulatory system. Increasing in size, organisms such as crustaceans (crabs) and some bivalve mollusks (clams) have semi-open circulatory system which comprise some veins and arteries, without capillaries. This followed by still larger animals that have a closed circulatory system. Within the group that has a closed circulatory system are found different types of hearts. For instance, the hearts of frogs have incomplete 3-chambered hearts, reptiles' hearts have three chambers, crocodiles have incomplete 4-chambered hearts and the hearts of birds and mammals have four chambers.

LIVING **ARTHROPODS**	STRUCTURE OF CIRCULATORY SYSTEM	BODY SIZE
Eriophyidae (Acari)	On the whole, lack the circulatory system. Lack the vessels and a simplex heart.	Very small and microscopic
Insects	Open circulatory system,	Small
Crabs	Semi-open circulatory system has some arteries, without veins and capillaries.	Largest species among Arthropods

LIVING **WORMS**	STRUCTURE OF CIRCULATORY SYSTEM	BODY SIZE
Nematoda Bryozoa	On the whole, lack the circulatory system	Very small and microscopic
Phoronidea Sipunculoidea Brachiopoda	These worms have a simple circulatory system consisting of some linear vessels and a simplex heart	Little larger
Segmented worms	These worms have an advanced circulatory system, consisting of linear and partial vessels, some hearts and capillaries. Their blood circulation is closed.	Largest species among worms. Rhinodrilus Fafneri which reaches 210cm long and 2.5cm diameter, and Eunice Gigantea which is 3m, long.

LIVING **MOLLUSCA**	STRUCTURE OF CIRCULATORY SYSTEM	BODY SIZE
Amphineura Gastropoda	Open circulatory system, few vessels and a simplex heart	Small
Bivalves	Circulatory system is more advanced. Semi-open circulatory system has some veins and arteries, withot capillaeies.	Large size
Cephalopoda	Chephalopoda, are the most developed ones of Mollusca, and have a closed circulatory system with many vessels and heart chambres	Largest speces among Mollusca Octopuses and Architeuths, which have a body of 2.5m long and arms as longs 12-18m.

LIVING LAND VERTEBRATE	STRUCTURE OF CIRCULATORY SYSTEM	BODY SIZE
Amphibians	Larvas have the 2-chambered hearts. Larva usually aquatic. Adults have the incomplete 3-chambered hearts.	small
Reptiles	Circulatory system is imperfect and their dark blood is mixed together with the bright one, because their heart is 3-chambered. An incomplete wall between their ventricle.	Reptiles can not hold their head upward, and have to creep on the ground and have shorter legs. Little larger
Crocodiles	Incomplete 4-chambered hearts. An complete wall between their ventricle, but their dark blood is mixed together with the bright one.	Larger of reptiles
Birds	After mammals, birds have the most developed circulatory system. Their heart is also 4-chambered	Larger, whit long necks and legs
Mammals	Mammals have the most developed circulatory system and 4-chambered heart.	Largest land animals. Mammals have the longest necks and legs.

Zoologists put all animals in the **three** groups:

1) Lack the circulatory system
2) Open circulatory system
3) Closed circulatory system

But **four** the first time, I use the term **"semi-open circulatory system"** and put the animals in four groups.

1) Lack the circulatory system
2) Open circulatory system
3) Semi-open circulatory system
4) Closed circulatory system

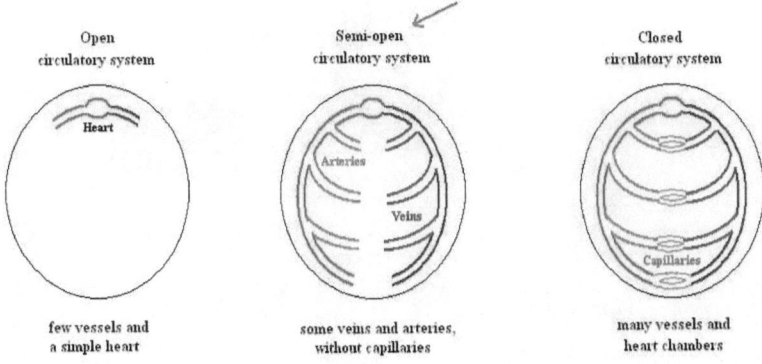

Open circulatory system	Semi-open circulatory system	Closed circulatory system
few vessels and a simple heart	some veins and arteries, without capillaries	many vessels and heart chambers

Ramin Aminnardfar
1 April 2015

I have found this matter for the first time that **Crocodiles are not a reptile**. Crocodile such as birds and mammals are a separate class, which have been derived from reptiles.

Mammalian four-chambered heart from a reptile-like three-chambered heart is extension of the septum (the wall dividing the chambers, lost of the right systemic arch and the left systemic arch persists).

In **birds**, lost of the left systemic arch and the right systemic arch persist.

The lineage leading to **crocodiles** evolved a four-chambered heart along a different pathway, keeping both systemic arches.

Zoologists putting the land vertebrates in 4 classes (amphibians, reptiles, birds and mammals). But I came to the conclusion that their crocodile are a separate class. As birds and mammals, with obtaining, four-chambered heart, have been derived from the class of reptiles and create a separate class. Crocodile also, have made four-chambered hearts and create a new class and has been derived from the reptile class. Mammalian four-chambered heart from a reptile-like three-chambered heart is extension of the septum (the wall dividing the chambers, lost of the right systemic arch and the left systemic arch persists). Vice versa in birds, lost of the left systemic arch and the right systemic arch persist. The lineage leading to crocodiles evolved a four-chambered heart along a different pathway, keeping both systemic arches. We should put the land vertebrates in 5 classes (amphibians, reptiles, crocodiles, birds and mammals).

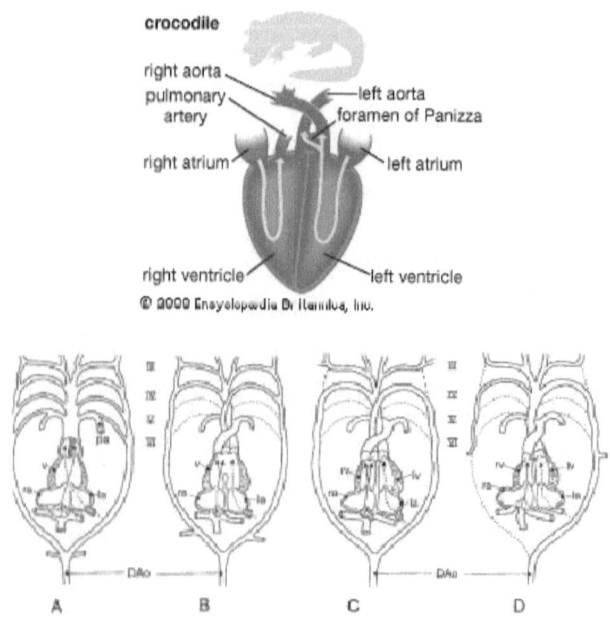

The lineage leading to crocodiles evolved a four-chambered heart along a different pathway than mammals, keeping both systemic arches.

4.Relationship between the power of the blood/fluid circulatory system and size of plants.

In plant's bodies, the root, which is like the heart of heart of animals, takes the water and other materials upwards, vascular system which are like blood vessels in animals, lead the water and materials to upper organs of plants. As we saw in animals, the evolution of transferring system in plants is not at the same degree. Among the plants there are primitive and evoluted classes. There is a question that; is the direct relationship between the strength of transferring system and the size of body, true in plants too? To answer the question, it is better to classify the degree of the evolution of transferring system of today's plant classes. The group of angiosperms has the most advanced transferring system. Then there are gymnosperms which have more advanced transferring system. Then ferns have relatively strong root and vascular system. The group of horsetails has weak and primary roots and vascular system, then there are lycopodiales, which have the most primary roots and vascular system, and at the end there are bryophyta which almost lack this system.

Now, let's compare the size of body between these groups. All plants which we see as tall trees in forests and other parts of earth, belong to the groups of angiosperms and gymnosperms that is they have strong transferring system! Ferns are next group, which have large sizes. Horsetails are able only to reach to 1 meter and lycopodials are smaller than they are, and at the end there are bryophyta, which grow crawling on the earth, or on other plants. So, we see that the direct relationship between the strength of transferring system and the size of body is true about the plants. This is a normal situation, because, it needs to overcome the gravity to transfer the water and materials to high parts and leaves of plants, and it is clear that, the stranger the pump system and leading system, the more the distance of transmission, so the plant will be able to transfer the water and materials to high parts, and if this system week, the plant had to shorten is height

LIVING LAND PLANTS	STRUCTURE OF FLUID CIRCULATION SYSTEM	BODY SIZE
Bryophyta (Moss)	Almost lack fluid circulation system	**Very small**
Lycophyta (Club moss)	Most primary roots and vascular system	**small**
Sphenophyta (Horsetails)	Weak and primary roots and vascular system	**smallish**
Ptreophya (Ferns)	Relatively strong root and vascular system	**largish**
Gymnosperms	Advanced fluid circulation system	**Large**
Angiosperms	Advanced fluid circulation system	**Large**

Just as animals have to combat gravity to survive and reproduce so do plants and once again one can witness a relationship between the size of organisms and gravity. One saw the problems that the giraffe faced in the supply of blood to the brine and body extremities. The problems is the same for plants, tall plants require water and nutrients transferred to all parts including their most elevated leaves and branches. The taller the plant the more efficient the transfer system of fluids needs to be. What is interesting is the plants have evolved with different structures that facilitate growth and height. Similar to the evolution of the animal heart, plants being their evolutionary origins with the most rudimentary system of fluid distribution. The first colonization of terrestrial earth 510-439 mya by plants was with bryophytes (liverworts and mosses). Bryophytes still exist today but with only a most rudimentary vascular system to transfer fluids and their size is constrained by gravity to a maximum height of one meter. They also lack characteristic stems, leaves, or root, and adhere to the ground by the use of rhizoids, which are only small and hair like. See Table to help understand the vascular system.

1. Dose the force of gravity increase?

As we know, there is a direct relationship between the power of a circulatory (blood) system of an animal and its size. In animals, the heart tends to push the blood towards the top of the body where the head is located while gravity force pulls the blood down and tries to prevent it from rising. Thus the heart must overcome the force of gravity in order to push the blood towards the top of the body. The more it is successful, the more the blood rises and the height of the animal is allowed to increase. Mammals are the largest animals because they have the most developed circulatory system and, more than other animals, their 4-chambered heart can overcome the force of gravity and send the blood upwards towards the head where the brain is located. For this reason, mammals have the longest necks and legs. After mammals, birds have the most developed circulatory system. Their heart is also 4-chambered and, to a large extent, it can overcome the force of gravity and also send the blood upwards towards the top of the body, where the head of an ostrich, for example, and other tall birds is located. This is the reason why the birds' necks and legs are also rather long. The blood circulation system, both in mammals and birds, are perfectly segregated and the bright blood is separated from the dark. The reptiles are in the class lower than mammals and birds. Their circulatory system is imperfect and their dark blood is mixed together with the bright one, because their heart is 3-chambered. The 3-chambered heart can send blood forward and this is the reason why reptiles can not hold their head upward, and have to creep on the ground in the lying position and have shorter legs, and we call this action creeping, and the animals are called reptiles. So we observe that because of lack of a strong heart, they have to creep on the ground, for their heart cannot transmit much more blood from the body and send it upwards. However, reptiles' heart, contrary to amphibian's, is much more effective because they have an incomplete wall between their ventricle, although it cannot send the blood upwards, at least it can make it move forward horizontally, and because of this, the reptile's body, such as Lizards, Crocodiles and Snakes, grow horizontally. This is not the case for the amphibians and their body, which is smaller, both horizontally and vertically. Although the fishes' hearts are 2-chambered, they can grow horizontally because they have a perfect circulation system, not mixing their dark blood with the bright one, and being aquatic, because horizontal movement of blood does not need to overcome gravity and their heart is able do this. Understanding the relationship of the strength of the heart and the body size and design of animals is easier when we study, in depth, the living animals we have on the earth today. But in the past, there were animals on the Earth, as well as the present animals, which do not exist today, and we discover them by their fossil traces. Such as with today's animals, was there any relationship between the strength of blood circulation system of animals in the past and the size of their body? Dinosaurs are the primary animals of the past that one might consider, because their huge body is intriguing and holds an interest for everyone. The dinosaurs belonged to the reptile class and had a 3-chambered heart. Thus, the old reptiles, with a 3-chambered heart, and an incomplete blood circulation system, could grow so large and hold their heads upward, and have long necks and legs. Where as today's reptiles cannot even raise their head even slightly, and have very short legs, or do not have any legs at all, and must creep upon the ground completely. Isn't this strange? Isn't the relation of strength of the heart and the size and design of an animal's body also supposed to be true for the dinosaurs? It's thus perhaps better to study about other old animals and compare them, in terms of strength of the heart and body size, in that period, to find the reason why dinosaurs, having an incomplete circulatory system, could grow so large. If we study more carefully, different animal species in the past, we can perhaps arrive at the following results: 1. Observing this phenomenon, from past to the present, the body size of all animals has been gradually decreased. In the Mollusks class, there was an animal 4.5m in diameter called Endocras, which their equals today are only a few centimeters in thickness. In the insect class, there were dragonfiles (Meganeuropsis) 71cm long, which their equals today are much less than 15cm. In the amphibian class there was an animal called Eogyrinus, 4.5m long, but today's amphibians are not more than few centimeters. In the reptile class, Brontotosarus and Tyrannosaurus existed in the past but now alligators and other Crocodiles are the largest reptiles. From the mammals in the past, the Mammoth, Mastodon, and the largest mammals, that is, Baluchithere and also the Giant camel existed - but today they do not. 2. In each period of time, there has been the direct relationship between the strength of the heart and the size of the body. In the past 500 million years, Endocras was larger than all other animals because its heart was stronger than all of them and the blood circulation system were more developed than were the others. In the past 330 million years, when the amphibian Eogyrinus (with 4.5m long) and Erypos with 2.5m long existed in the Carbonipher period, they were larger than all the animals living on the land, because they had the strongest heart of all the animals at that time. Then in the past 200 million years, reptiles became the largest, because they had the strongest heart at the time. Mammals like Baluchithere and Mammoths became the largest land animals on earth in the past 20 million years, because at that time they also had the strongest heart. Nowadays, mammals such as elephants and Giraffes have the largest bodies among land animals because they have the strongest heart of all the animals today. 3. The third observation is that as we close to the present time, from past to today, animals need stronger hearts to create larger bodies. In the Carbonipher period, the heart of an amphibian could allow an animal to reach 4.5m long, but the same heart today only allows an animal to become a few centimeters high. In the past 200 million years, a 3-chambered heart of a reptile could create a dinosaur, but now it can only create animals as large as Snakes and Lizards. In the past 20 million years, a 4-chambered heart of a mammal could allow for a Baluchithere and other large mammals, but today the same heart can only create animals as large as elephants and giraffes. (Although the Whale is larger than the elephant and giraffe, but this characteristic is because of it being aquatic and because its body grows in a horizontal state, and if the whale lived on land like the elephant and tried to raise its body vertically and stand, it wouldn't be larger than the elephant and giraffe.) In the past 250 million years, the blood circulation system in insects, could create dragonfiles 71cm long, but now it can only allow for 15cm ones. So far, we know that the heart needs to overcome the gravity force of the Earth in order to send blood upwards to the heads of animals. We observe that any animal with strong heart can overcome this force better and this contributes to make the animal higher and taller. But how is it possible that an animal with specific heart constraints at one time, can overcome the gravity and become larger while in at a later time, the same specific heart constraints cause the animals to have a smaller body? Only in one case this is possible and that is if we suppose the strength of gravity at these two times is different. We can thus justify why it is in the past that an animal with an identical heart to today's animals is larger than its modern counterpart. Such as it is, that the large animals did not extinct suddenly, but rather gradually, so it must be that the increase of gravity from past to present has also occurred gradually. When a Molluscan Endocras, that had an open blood circulation system, in the past 500 million years, could reach 4.5m in diameter, and move its large shell and keep on living, certainly the gravity must has been lower in order to allow for such a result. From that time to the present, no shell has been created because the gravity has increased and this animal could not grow it in these conditions. In the past 330 million years, when an amphibian, 4.5m long, existed, its 3-chambered heart and incomplete blood circulation system could easily overcome the gravity and send its blood upwards to its head. Surely at that time, the force of gravity must have been lower than it is today. Since then, no amphibian has become so large because the gravity increased and the amphibian's heart could no longer overcome the force of gravity, which caused their bodies to become small. We can thus cite a primary reason for the extinction of the dinosaurs, if we accept that gravity has increased from the past to the present. It is interesting that the dinosaurs themselves did not become extinct suddenly, and that their extinction has come about gradually, throughout millions years. An additional, most important issue is that their extinction began from the larger animals, because the gravity increased gradually, and prevented their blood from reaching their head, and as a result, the animals had to gradually decrease in height and body size. In other words, the larger animals became extinct and the smaller ones survived, and only counterparts who are small and who creep on the ground were amongst the survivors. The animals in the mammal class have also lost many of their largest types, such as Mammoths, because of gravity increase. An interesting point is that presently, the larger types of this class are in danger of extinction; that is, the elephants and the rhinoceroses. From this we might conclude that the continuing increase in gravity persists and that it is bringing an end to the taller and larger animals. We may not be able to perceive this so clearly today, because this increase in gravity occurs very slowly, and its effects only become evident through much longer time spans, such as thousands and millions of years. .

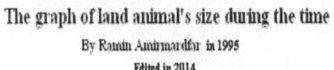

The graph of land animal's size during the time

By Ramin Amirmardfar in 1995

Edited in 2014

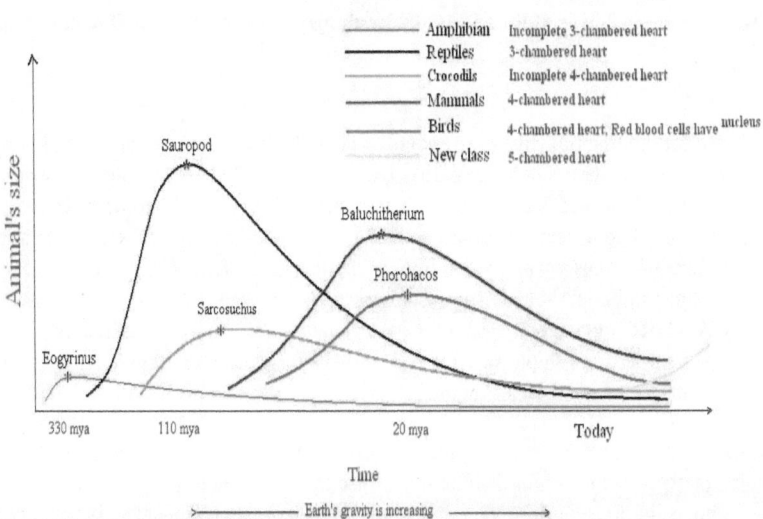

The figure above shows reasoning about animal's sizes animals that have a stronger blood circulatory system could have larger size Plants that have a stronger fluid circulatory system could have larger size. If an animal's/plant's blood/fluid circulatory system is weak, it definitely has a small size. If an animal's/plant's size is large, it definitely has a powerful blood/fluid circulatory system. But if its blood/fluid circulatory system is powerful, its body is not necessarily large! And if its bulk is small, its blood/fluid circulatory system is not necessarily weak! Animals and plants size evolution during the Geological Time. Graph of the land animals' size since ≈350 Ma ago to the Recent. Graph of the land plants' size since 400 Ma ago to the Recent and Future. "Y" axis represents body size of the biggest species in each category at a specified time. "X" axis is the indicator of the time or indicates the amount of gravity at any time. Fall of the diagrams (decreasing slope) indicates that body size of species are getting smaller while the gravity is increasing. Force of the gravity on the blood/fluid circulatory system will prevail. Climbing of the diagrams (increasing slope) indicates that species are getting bigger as the result of evolution and as blood / fluid circulation system gets powerful. The force of blood / fluid circulation system will overcome the force of gravity. If the circulatory system of blood /fluid would not been evolved, diagrams had had only a decreasing slope and all animals and plants, under the effects of increased gravity, were small millions of years ago and now we couldn't see any big animals and plants. Evolution of an animal or plant is possible under the effect of changing environmental conditions and selection of more adapted individuals to new situations by natural selection. The main factor driving the evolution of circulatory system is increasing of the gravity. Means, if the gravity had not been increased, the blood and material circulatory system was not needed for development. An intriguing prediction of future evolutionary process of animals and plants can be hypothesized according to these diagrams (new class and the type of their blood/ fluid circulation system). "New class" will evolve out of small mammals. The new class will have (4+1)-chambered heart. (4+1)-chambered heart = A four-chambered heart with a single-chambered heart (Subsidiary heart). Preparations this heart has been developed inside the body of modern mammals.

2. Relation Between Power of Blood/Fluid Circulatory System and Size of Animals and Plants

I have argued for the first time that each animal or plant that has a stronger blood/ fluid circulatory system could have a larger bulk (Fig 1a); that animals with no circulatory system such as nematodes and plant's mites, have a tiny body size. Animals that have an open circulatory system (without arteries, without veins, without capillaries) – such as insects and some mollusks– have small body size. Animals owing a semi-open circulatory system (Comprise some veins and arteries, without capillaries) – such as crustaceans and some mollusks (bivalve) – have average body sizes. A closed circulatory system (Comprise arteries, veins and capillaries) is associated to animals – such as some mollusks (Octopus), Annelid worms (earthworms) and vertebrates (fish, amphibians, reptiles, birds and mammals) – having a large body size.

Zoologists put all animals in the three groups: no circulatory system, open circulatory system and closed circulatory system. But for the first time, I use the term "semi-open circulatory system" and put the animals in four groups: no circulatory system, open circulatory system, semi-open circulatory system and closed circulatory system.

I begin with an analogy. Let us consider a city and its population. Each individual lives in a large or a small house depending on his/her financial condition. This means that if someone is well-to-do, he probably would live in a spacious house, whereas someone less well-off is not able to afford a roomy house and therefore is constrained to live in a cramped one. It is common sense and everyone accepts this easily. Similarly, during my inquiry I found out that animals and plants having a stronger heart or fluid circulatory system could afford larger body masses

This is because their strong system enables them to better overcome gravity and to pump fluids to greater distances and to higher levels. It is a matter of logic that everyone can understand and accept. But if this is so obvious, why are scientists not aware of it? What is the obstacle to find out the relation? Why is it that this simple relation between the power of blood/fluid circulatory system and bulk size of animals and plants has been overlooked? Where lies the problem? To find this out, let us go back to our example and ask ourselves some questions. Do people who are in a poor financial state live in small houses? The answer is "Yes", because no one who disposes of few money is able to buy a large house. Do all people who are well-to-do live in large houses? The answer is "No"! This is because some who have a lot of money live in small houses for whatever reason. It means that some of the rich may live in large houses, while others inhabit a small house, without any restriction to them. So, when we look at the houses of a city and see large houses among them, we can definitely say that their owners are rich. But when we see a small house, we cannot say that their owners are poor. Because they may be rich!

In the biological world things are similar. If the blood/fluid system of a living organism is weak, it is forced to have a small body (Fig. 1c). However, if its blood/fluid system is powerful, the body must not necessarily be large! It may be large or small, depending on different other factors (Fig.1e). But if it is large, it means that the circulatory system is strong (Fig. 1d). We can thus conclude that small animals or plants may have weak or strong blood/fluid systems (Fig. 1f). This confusing fact is the issue that has caused scientists not to find the relation between the power of blood/fluid circulatory system and bulk size of animals and plants.

To solve the problem, we should at first classify animals and plants according to their power of blood/fluid circulatory systems, then choose the largest species of each group and made the comparison among them. We should not involve animals having strong heart/small size in our consideration and compare only between animals having weak heart/small bulk and animals having strong heart/large bulk. By proceeding this way, the direct relation between the power of blood/fluid circulatory system and animals/plants bulk size will become evident.

Now, what about those "rich people living in a small house" – the animals that have strong hearts but small sizes? We should find a reason for their existence. We should ask ourselves what is the factor preventing them to live in larger houses. The animal has strong heart but small size. What is the factor preventing it to get larger? Scientists, instead of following the causal chain of the process, facing the difficulty of the job, become disappointed and discouraged to find the solution and do not ask to themselves the following correct questions:

A) Why have some of the animals and plants small sizes while they have strong blood/fluid circulatory systems? What is the factor preventing them to get larger?

B) Why the nature, have put such a factor inside the body of animals and plants so that their bulk sizes remain small despite of having strong circulatory system? Why the nature needs such animals (strong heart and small bulk)?

These two questions have been discussed and answered in two chapters of the book (Mardfar, 2000) and are two of the most important chapters of my book. It is impossible to create a logical and simple relation between blood/fluid circulatory system and body size of animals and plants without finding the answers of these two question and that why scientists previous to me were not able to find the relation.

3. Evolutive Importance of the Circulatory System/Gravity Interaction

Rotation of Earth causes a centrifugal force on things, which has a component on opposite direction with respect to gravity and causes a weight decrease of them. This force is stronger on the equator, so weight of a thing at 0∘N is a little bit lighter than the weight of the same thing on other latitudes of the Earth. Centrifugal force causes that while a thing has 1000gr weight on the pole, it will be 996gr on the equator. In other words, we see about 4gr decrease per kilogram. A person with 70 kg at the pole, will be 69.7 kg at the equator. If we transfer a thing with 1 kg from latitude 45∘ to the equator, difference of weigh per kg will be less than 4 gr. Therefore we see that little change of gravity on the Earth causes an insignificant difference in thing's weight.

In 1671 an astronomer group set off from Paris (45∘N) to Cayenne Island (5∘N) for a commission. Jean Richer, the head of this group, had a precise pendulum clock with him in this trip. He noticed that clock was postponed 2.8 minutes per 24 hours. Jean Richer did not know why. But about 15 years later, Isaac Newton said that the reason is the difference of gravity in the mentioned zone. After that, pendulum became a precise facility for measuring difference of gravity in various points of the Earth. Because in places near together, spring balance could not show insignificant difference of gravity but pendulum could easily. It seems that pendulum can acts like a magnifying glass. How a pendulum can do such a work? When we weigh a thing with a spring-balance, and compare the difference of weight with other zones, we only have done and resisted this job once, therefore the difference will be little and insignificant. Analogously, if pendulum move only once, the difference of time will be very little; i.e. the difference of time for a single oscillation of pendulum on equator and pole is insignificant and shows the same little difference of the spring balance. But when pendulum moves twice (goes and returns) it will delay a little each time and, because of gathering together of these delays, we will see the difference twice larger i.e. pendulum has magnified it twice, so we can see it easier. If pendulum moves three times, the difference will be three-fold. So it is possible to compare each other the difference in gravity on various zones. Imagine if pendulum go and re- turn once per second, it has many times to go and return in 24 hours and it is possible to obtain a good magnification. This great magnification was that Jean Richer ascribed in 1671 to the difference in gravity between Paris and Cayenne Island. If he wanted to do this test even with the most precise balance of that time, he would had not get a conclusion and could had not put in evidence any difference of gravity between Cayenne island and Paris.

There is a apparatus similar to a pendulum in animal's body, which shows the little difference of gravity with a great magnification. This apparatus is the same blood circulation system with the heart engine.

Blood moves like pendulum cyclically. Each tour of circulation of blood is similar to onetime round trip of a pendulum. Blood go aloft from the surface of Earth alternately like a pendulum and comes nearer to it again and repeats this duty hundreds of time each day and night. So, a little difference in gravity will have a lot of effect in work of this apparatus as well as it had in task of a pendulum.

If blood circulation was only once during the life of an animal, the little difference of gravity could not impress on it and acted as a spring balance. But because blood goes up and down from the surface of Earth thousands of times, so like a pendulum, the effect of little difference of gravity in this system will add up thousand times resulting in a great amount. If an animal is larger and have more blood, this additional load will be more effective.

In the following of this paper the relation between blood/fluid circulatory system and size of animals and plants through geological time will be discussed. However, the possibility that latitudinal gradient of gravity can influence latitudinal selection of species in the Recent time deserves to be the subject of more deep and circumstantiated future researches.

The elephant heart weights 22 kilograms and circulates about 450 liters (720 kg) of blood. And its heart 28 beats per minute. The capacity of adult elephant stomach is 50 litters (80 kg)

If we transfer the elephant from the equator to a place near to the pole, elephant's blood will be heavier about 4gr per kg, 320 grams overweight per heartbeat. In other words, the elephant's heart will carry per heartbeat 320gr additional weight.

Additional weight per day = 320gr x 28 x 60 x 24 = 12902400gr

Additional weight per year = 12902400gr x 366 = 4722278400gr

The elephant live around 70 years

Additional weight per generation of elephant = 4722278400gr x 60 = 283336704000gr = 28,3336,704kgr

In other words, the elephant's heart must carry 283336704kgr more load than in the equator, per generation of elephant. Therefore we see that, with a much little change in gravity it be imposed a lot of additional force to the animal's heart.

This is why tall and big animals such as giraffe and elephant can only live in equatorial points of earth, because of less gravity in there (little amount) than other places and their heart can send blood to farther distances from the surface of earth. So in practice we see that much little increase of gravity has noteworthy effect in animal bulk/stature.

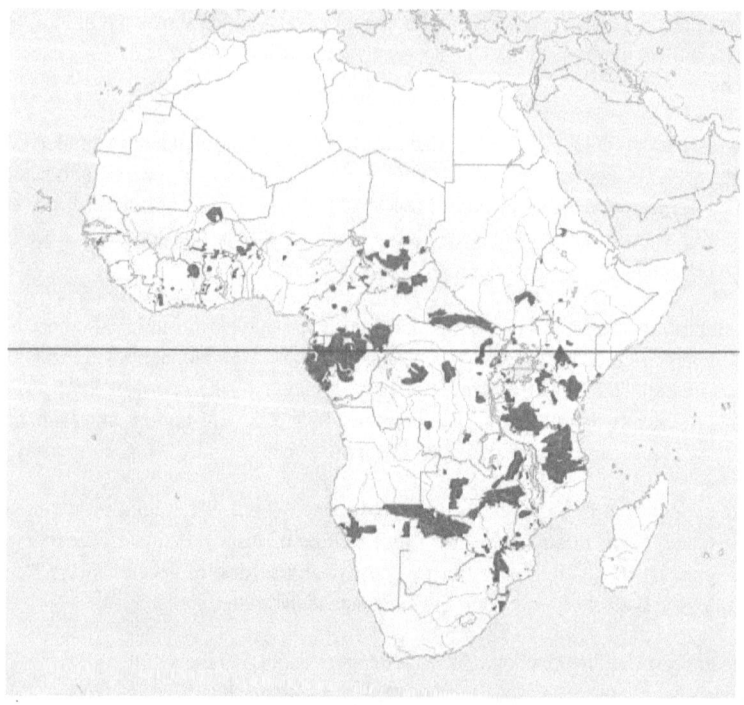

African Elephant (largest alive mammal) distribution map

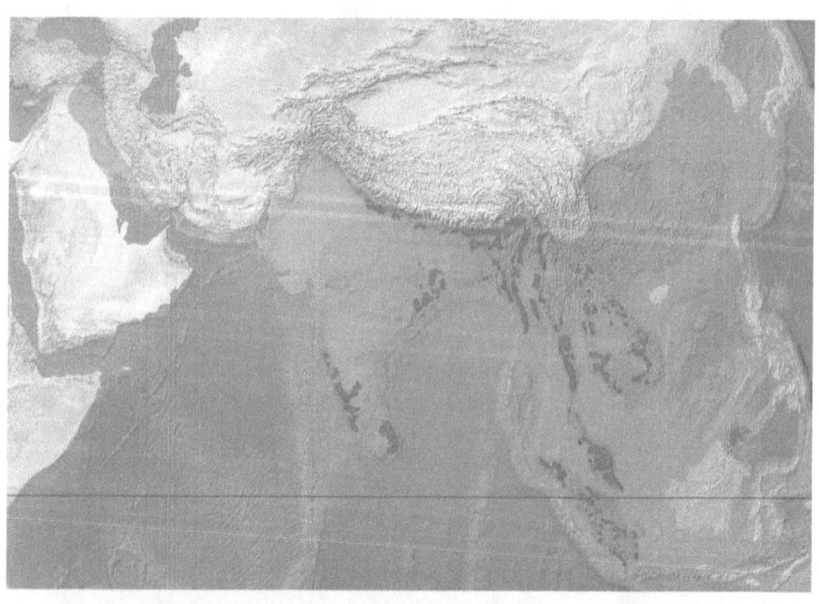

Elephas Maximus (Asian Elephas) distribution evolution map

4. Finding the Relationship Between Gravity and Evolution of Animals/Plants

If one day, a human can prevent him/her self from burning when alighting on the surface of the sun, under the effect of high gravity of sun, will stuck on the surface of sun like a sheet of paper. (Gamov, 1958)

I red this sentence in 1986 in a book titled Matter, Earth, and Sky written by George Gamow and concluded that live and death of animals are dependent on the STRENGHT of gravity. At first, I tried to establish a relationship by considering the skeleton and the amount of available food of the animals, but I faced with some obstacles. Later on I realized the importance of the circulatory system and thereby I explained on this basis the searched relationship. I must admit that I was lucky that taxonomist had classified animals/plants according to the power of circulatory system of blood/fluid. A "battle" between gravity and blood/fluid circulatory systems has always existed because most of the force of the circulatory system is used to overcome gravity. If gravity increases, animals/plants are forced to get smaller or to strengthen their blood/fluid circulatory systems. Nature has used both methods. Since evolution (strengthening the blood/fluid circulatory system) requires a long time, as gravity increases, animals/plants initially retreat and minimize their bulks. But when the evolutionary process succeeds in providing them a more powerful blood/fluid circulatory systems, animals and plants immediately move forward overcoming the gravity by maximizing their body sizes. Stage by stage new and stronger blood/fluid circulatory systems resulted by the evolutionary process, aiding animals to overcome gravity and to increase their size. But with further increase of gravity, the larger sized animals reduced their sizes again (minimizing their bulk) looking forward to evolve toward new and more powerful pumping apparatuses. If we draw the graph of changes of animal/plant sizes in the past we could witness the subsequent rise and fall of the bulk sizes of different classes of animals/plants in the past. The graphs have been constructed by:

1. Considering the classification of animals/plants due to their power of blood/fluid circulatory system.

2. Superposing in the same figure the curves of the distinct classes.

3. Not to involving "rich people living in small houses" in these figures (because they cause confusion).

These graphs clearly show the increasing gravity.

The graph of land animal's size during the time

By Ramin Amirmardfar in 1995

Edited in 2014

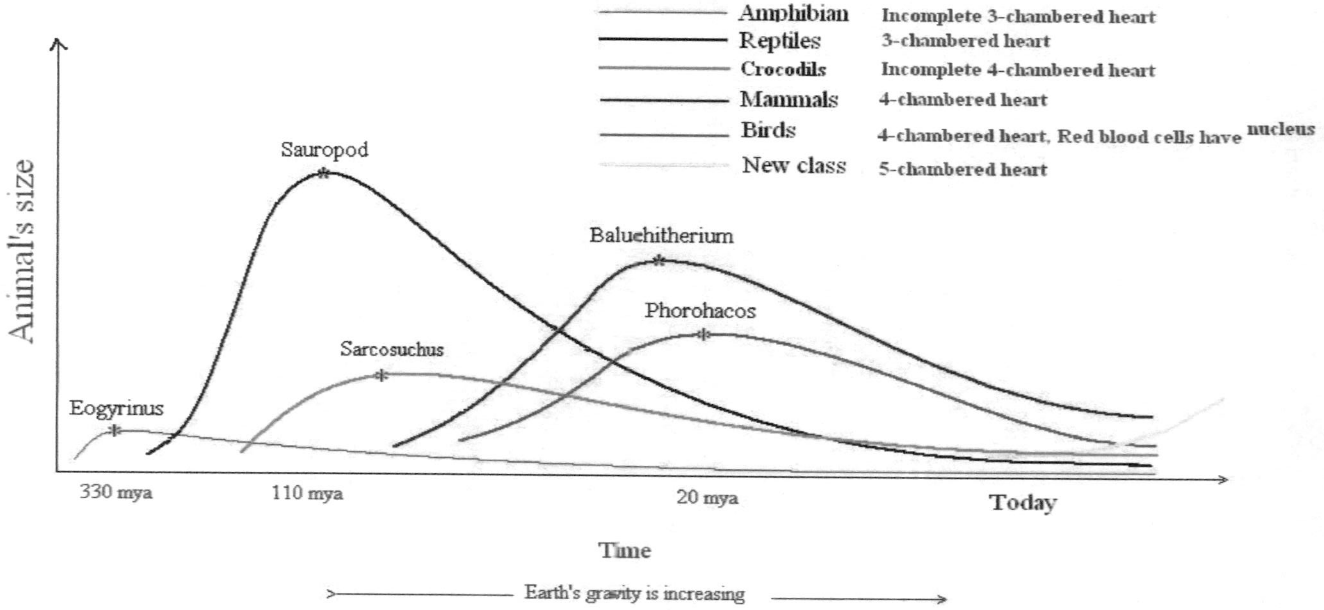

Animal's size

Time

Earth's gravity is increasing

330 mya

110 mya

20 mya

Today

Eogyrinus

Sauropod

Sarcosuchus

Baluchitherium

Phorohacos

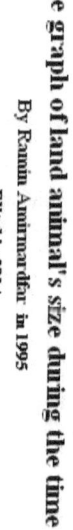

	Amphibian	Incomplete 3-chambered heart
	Reptiles	3-chambered heart
	Crocodils	Incomplete 4-chambered heart
	Mammals	4-chambered heart
	Birds	4-chambered heart, Red blood cells have nucleus
	New class	5-chambered heart

The graph of the land animals' size since ≈350 Ma ago to the Recent and Future. – b) Graph of the land plants' size since 400 Ma ago to the Recent and Future. "Y" axis represents body size of the biggest species in each category at a specified time. "X" axis is the indicator of the time or indicates the amount of gravity at any time. Fall of the diagrams (decreasing slope) indicates that body size of species are getting smaller while the gravity is increasing. Force of the gravity on the blood/fluid circulatory system will prevail. Climbing of the diagrams (increasing slope) indicates that species are getting bigger as the result of evolution and as blood / fluid circulation system gets powerful. The force of blood / fluid circulation system will overcome the force of gravity. If the circulatory system of blood /fluid would not been evolved, diagrams had had only a decreasing slope and all animals and plants, under the effects of increased gravity, were small millions of years ago and now we couldn't see any big animals and plants. Evolution of an animal or plant is possible under the effect of changing environmental conditions and selection of more adapted individuals to new situations by natural selection. The main factor driving the evolution of circulatory system is increasing of the gravity. Means, if the gravity had not been increased, the blood and material circulatory system was not needed for development. An intriguing prediction of future evolutionary process of animals and plants can be apotheosized according to these diagrams (new class and the type of their blood/ fluid circulation system). "New class" will evolve out of small mammals. The new class will have (4+1)-chambered heart. (4+1)-chambered heart = A four-chambered heart with a single-chambered heart (Subsidiary heart). Preparations this heart has been developed inside the body of modern mammals.

To discover the rate of gravity changes from past to now. Comparison of a dinosaur with a modern elephant is a wrong comparison, and data can not be true. Compare a dinosaur with a today tree on the grounds that both are organism, is this a correct comparison? Do we can find with this comparison, the correct data about changes in gravity from past to the present? You compare, a dinosaur with a modern elephant, on the grounds that both are animals. The two are different from each other, just as different from a tree with a dinosaur. A dinosaur should be compared with its counterpart in today. Dinosaurs are belonging to class of reptiles. Counterparts of dinosaurs are today reptiles. To discover true rate of gravity changes from past to now, you should compare the dinosaur with the largest living reptile. An elephant is not a reptile. An elephant physiology is very different from a reptile (dinosaur). Comparisons should be at least within the classes. If the comparison be within a order or a family, the data are more realistic. If the comparison be within an order or a family, the data will be more realistic. In this regard, see diagrams below.

Which of the following comparisons, will provide more true, data about the rate of change of gravity from the past to now?

Compare the two Earthly bodies

Ramin Amirmardfar
21 May 2016

Compare the two Organism

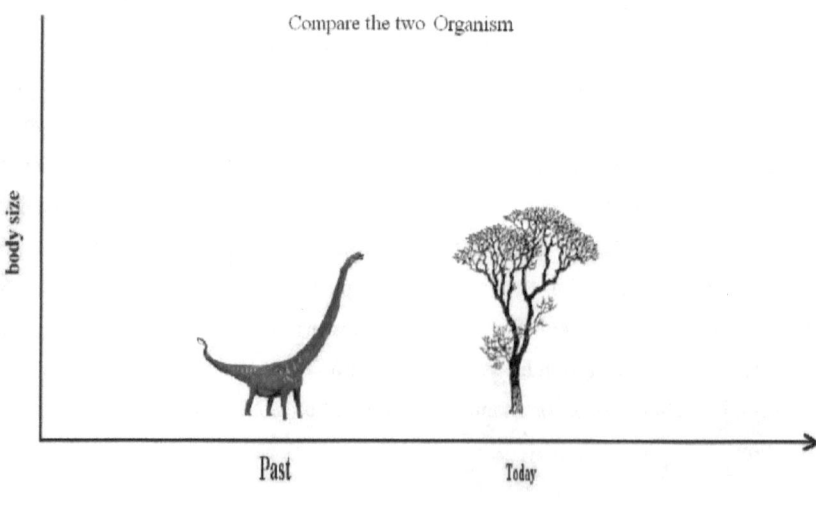

Ramin Amirmardfar
21 May 2016

Compare the two Animals

body size

Past   Today

------------- Earth's gravity changes ------------>

Ramin Amirmardfar
21 May 2016

Compare the two Reptiles

body size

Komodo Dragon

Past   Today

------------- Earth's gravity changes ------------>

Ramin Amirmardfar
21 May 2016

This is a wrong comparison. Because Dinosaurs are reptiles, but elephants are mammals.
The correct comparison is as below.

Change the size of reptiles

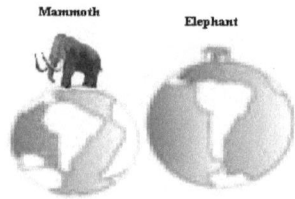

Change the size of Elephantidae

Change the size of Camelidae

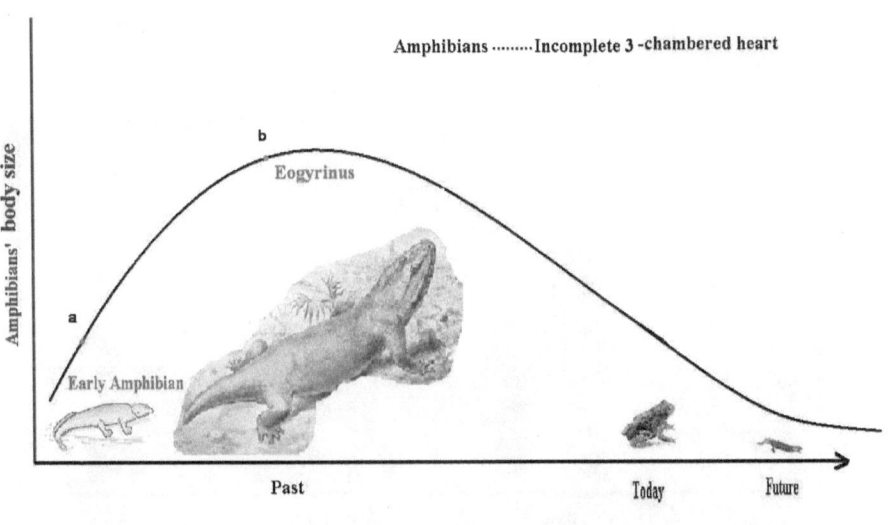

Amphibians ·········Incomplete 3 -chambered heart

Amphibians' body size

b

Eogyrinus

a

Early Amphibian

Past

Today

Future

·················· Earth's gravity is increasing ····················>

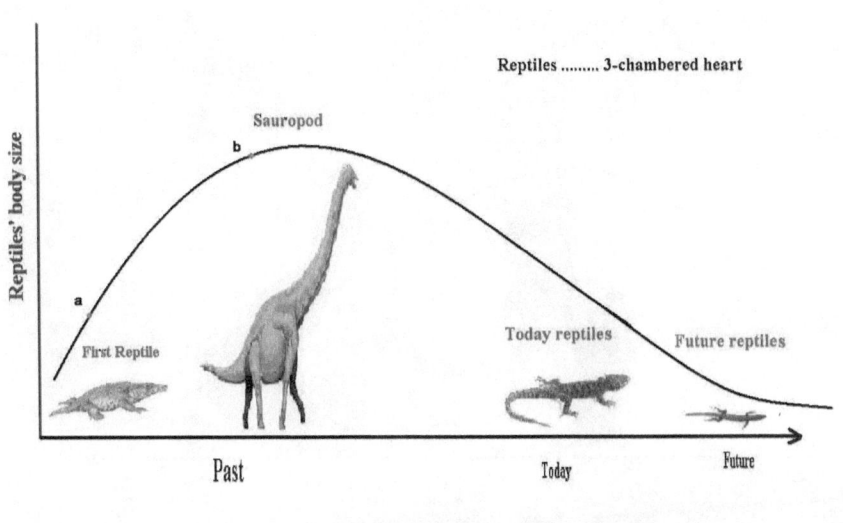

Reptiles ·········· 3-chambered heart

Reptiles' body size

Sauropod

b

a

First Reptile

Today reptiles

Future reptiles

Past

Today

Future

·············· Earth's gravity is increasing ·················>

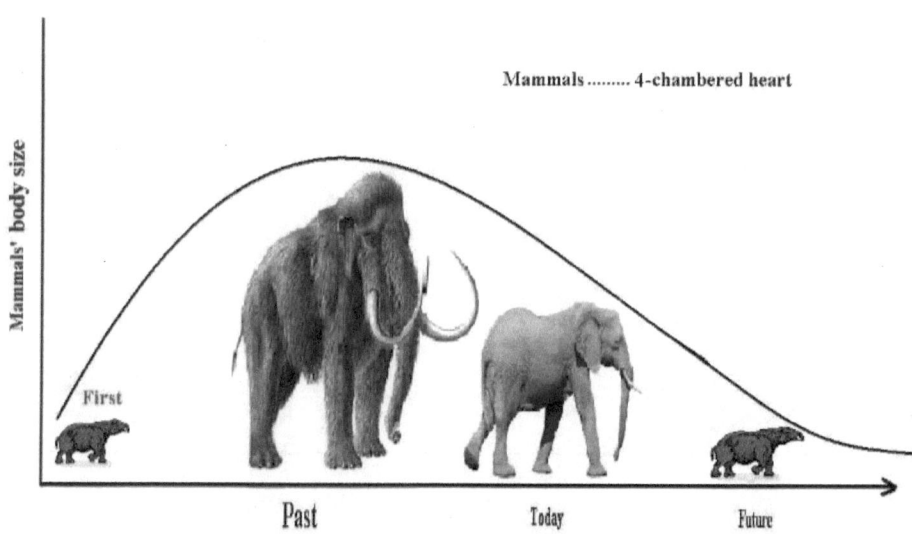

Mammals .... 4-chambered heart

Mammals' body size

First

Past

Today

Future

x---------------- Earth's gravity is increasing ----------------->

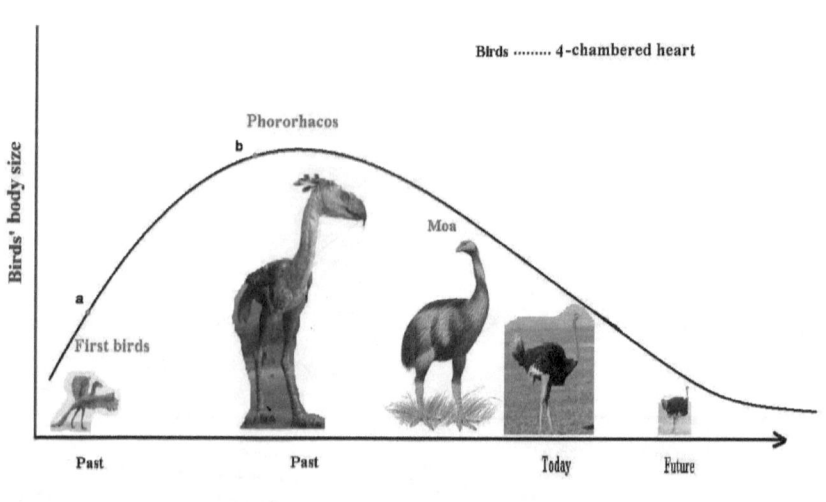

Birds .... 4-chambered heart

Birds' body size

Phororhacos

b

Moa

a

First birds

Past

Past

Today

Future

x---------------- Earth's gravity is increasing ----------------->

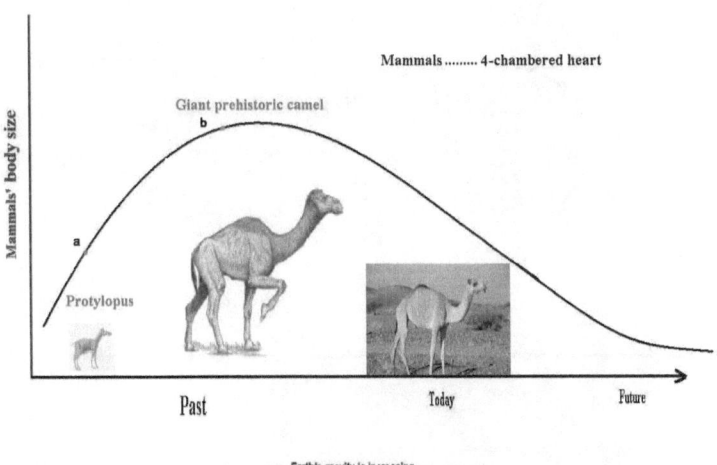

Mammals .... 4-chambered heart

Mammals' body size

Giant prehistoric camel

b

a

Protylopus

Past   Today   Future

>----------- Earth's gravity is increasing -----------

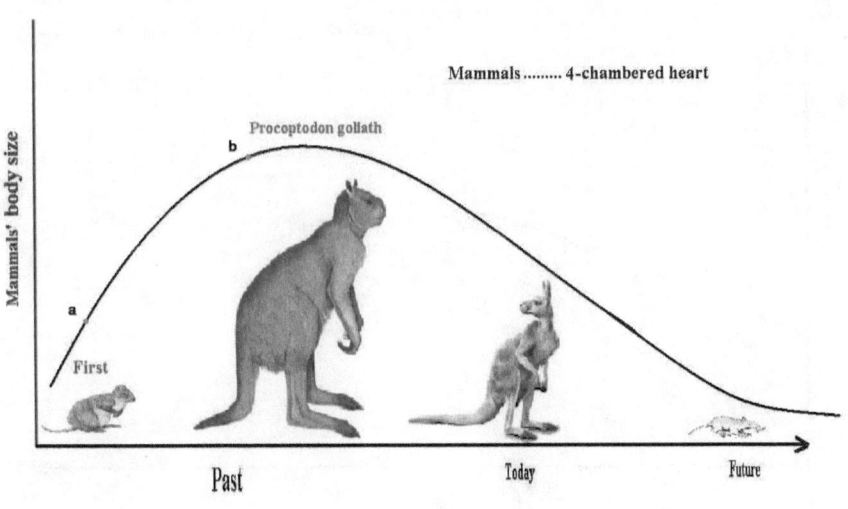

Mammals .... 4-chambered heart

Mammals' body size

Procoptodon goliath

b

a

First

Past   Today   Future

>----------- Earth's gravity is increasing -----------

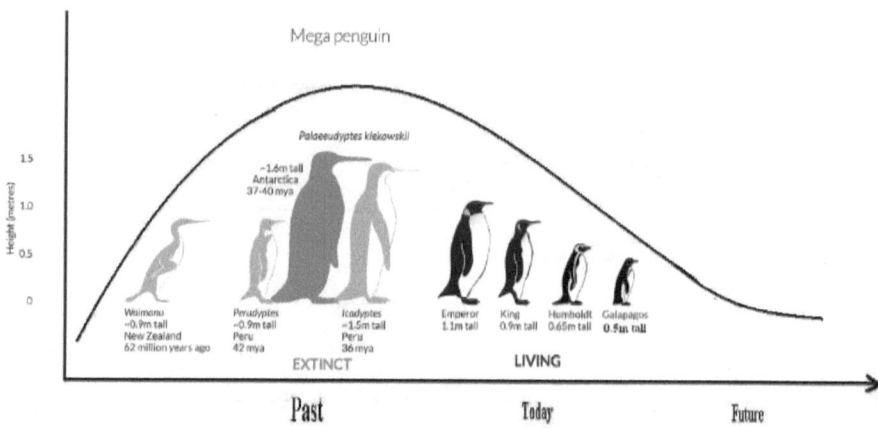

Ramin Amirmardfar
21 May 2016

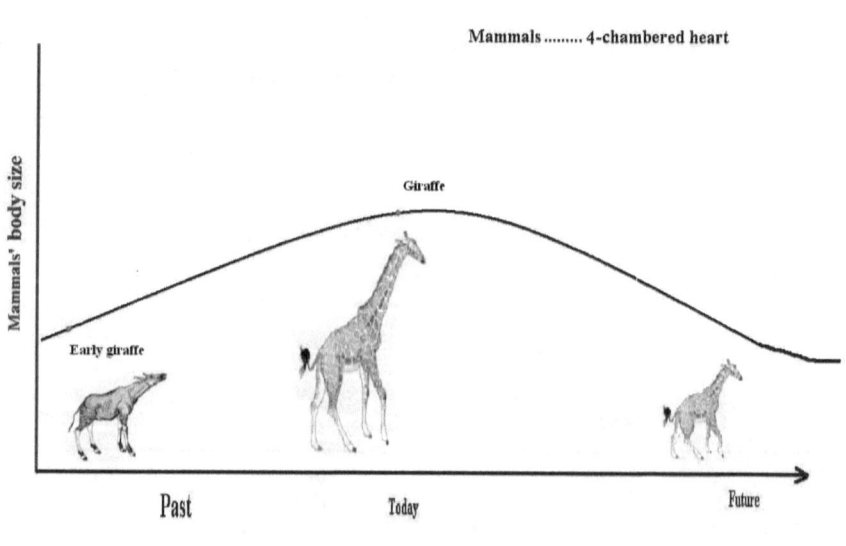

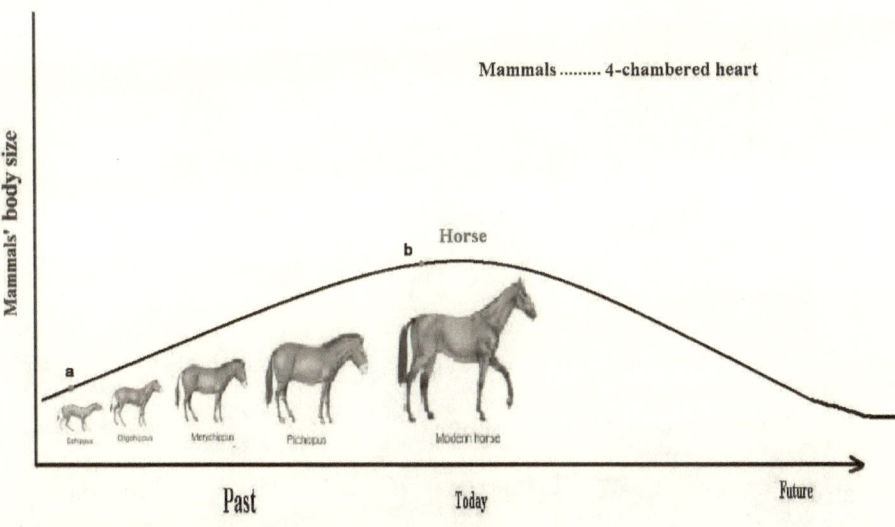

Mammals .... 4-chambered heart

Horse

Mammals' body size

Past   Today   Future

>------------- Earth's gravity is increasing -------------->

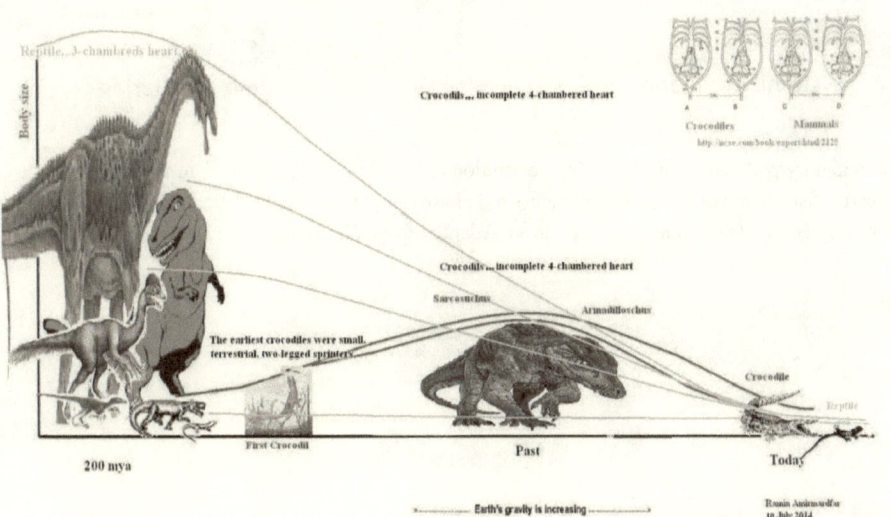

Reptile...3-chambereds heart

Body size

Crocodile... incomplete 4-chambered heart

Crocodiles   Mammals

http://acee.com/book/report/dated/2120

Crocodile... incomplete 4-chambered heart

Sarcosuchus

Armadilloschus

The earliest crocodiles were small, terrestrial, two-legged sprinters.

Crocodile

Reptile

200 mya   First Crocodil   Past   Today

>------------- Earth's gravity is increasing -------------->

Ramin Aminmadfar
10 July 2014

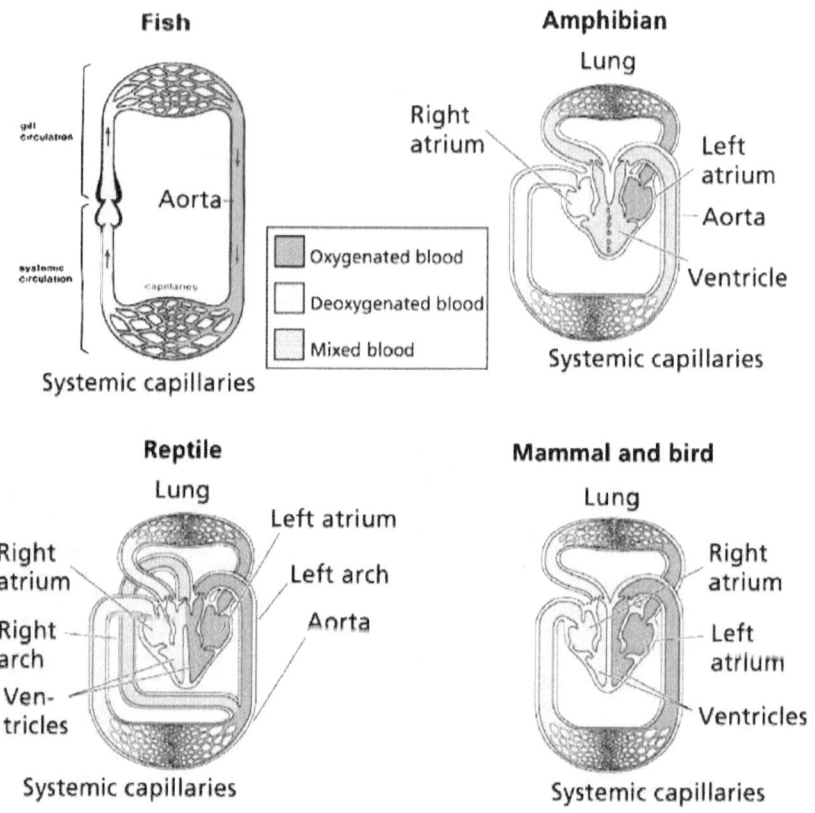

Circulatory paths in vertebrates. Heart evolution of vertebrates has begun from 2-chambered hearts (fishes) and after passing 3-chambered phase (amphibians and reptiles), has reached to 4-chambered hearts (in mammals and birds). Adapted from Wikispaces (2012).

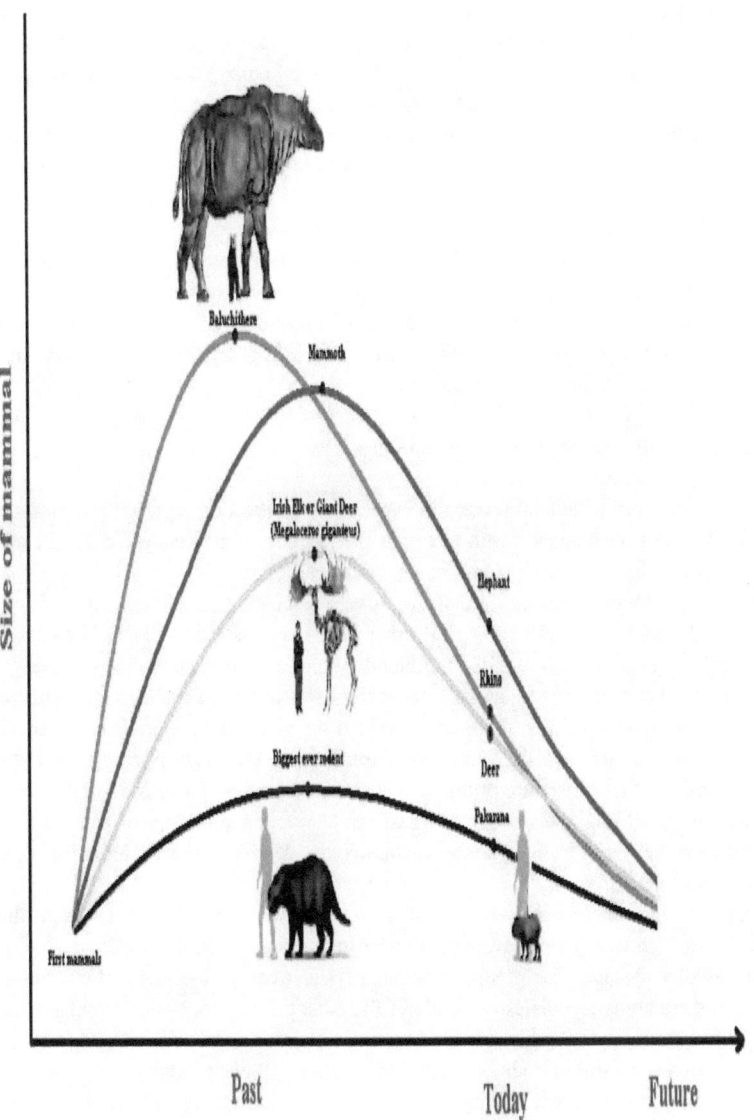

5. The evolution of mammals

Mammals evolved at the first time by small bulk animals. Mammals have managed to overcame gravity by the aid of 4-chambered hearts and strong blood circulatory systems and making their bulks larger (as shown on figures). But during the time a gradual increasing of gravity occurred, mammals failed against greater gravity and had to minimize their bulks. The process of minimizing the body size has continued up to now and will continue in the future Mammals will became gradually smaller and finally will have to crawl on the ground (like reptiles class that already had been forced to crawl). It is possible that in the future a new class of animals (among mammals) will occur having a stronger circulatory system that will be able to overcome gravity allowing the body size to evolve towards larger sizes again. If we compare a current dinosaur with a present elephant, it is a wrong comparison, because they belong to two separate classes. Also, if we compare the bulk size of a Baluchitherium with a Sauropod, such a comparison will not lead us to fact of increasing gravity. On the contrary, it will lead us the following: Reptiles class have been able to back the gravity more at their maximum point powerfulness, but mammals during their attack, have moved forward less than reptiles.

6. Plants, other witnesses for the increases of gravity

We knew that, the size of animal species in every class has become smaller from past up today, and we justified this phenomenon with theory of the increase of gravity. We know the other than animals; other creatures are living plant's range, on the earth.
If the assumptions about increase of gravity, are true, we most observe its effects on from past up to day. In order to transfer the water and other necessary materials, plants have a system to transfer these materials, which is a like the blood system of animals. In plant's bodies, the root, which is like the heart of heart of animals, takes the water and other materials upwards, vascular system which are like blood vessels in animals, lead the water and materials to upper organs of plants. As we saw in animals, the evolution of transferring system in plants is not at the same degree. Among the plants there are primitive and evoluted classes. There is a question that; is the direct relationship between the strength of transferring system and the size of body, true in plants too? To answer the question, it is better to classify the degree of the evolution of transferring system of today's plant classes.
The group of angiosperms has the most advanced transferring system. Then there are gymnosperms which have more advanced transferring system. Then ferns have relatively strong root and vascular system. The group of horsetails has weak and primary roots and vascular system, then there are lycopodiales, which have the most primary roots and vascular system, and at the end there are bryophyta which almost lack this system.
Now, let's compare the size of body between these groups. All plants which we see as tall trees in forests and other parts of earth, belong to the groups of angiosperms and gymnosperms that is they have strong transferring system! Ferns are next group, which have large sizes. Horsetails are able only to reach to 1 meter and lycopodials are smaller than they are, and at the end there are bryophyta, which grow crawling on the earth, or on other plants. So, we see that the direct relationship between the strength of transferring system and the size of body is true about the plants. This is a normal situation, because, it needs to overcome the gravity to transfer the water and materials to high parts and leaves of plants, and it is clear that, the stranger the pump system

and leading system, the more the distance of transmission, so the plant will be able to transfer the water and materials to high parts, and if this system week, the plant had to shorten is height

Living land plants	Structure of fluid circulation system	Body size
Bryophyta (Moss)	Almost lack fluid circulation system	Very small
Lycophyta (Club moss)	Most primary roóts and vascular system	small
Sphenophyta (Horsetails)	Weak and primary roots and vascular system	smallish
Ptreophya (Ferns)	Relatively strong root and vascular system	largish
Gymnosperms	Advanced fluid circulation system	Large
Angiosperms	Advanced fluid circulation system	Large

Could the plants, which have more primary transferring system and small sizes, larger their sizes in the past? If the answer is <<yes>>, we can hope that our assumption is true. So we study about the fossils of pervious plants. The remaining traces on the cools of carbonifer period (330 millions years ago), draws the scientific' attention. There are the traces of giant Lepidodendrals and giant kalamitals, the heights of some Lepidodendrals reach to 34.2m and the diameter of them reached to 1.8m, and the height of giant horsetails reached to 15m or more. And we also observe the traces of tree and tall ferns in these fossils. So it is clear that the answer of the question is <<yes>>, that is the pervious plants with weak transferring system, were able to larger their sizes, but they can't do this now, because the gravity was less than today, and they were able to overcome it. So the plants are other witnesses of gravity from past up to day.

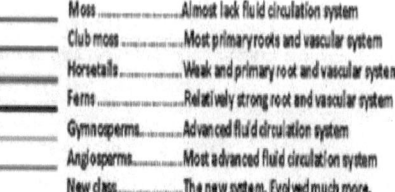

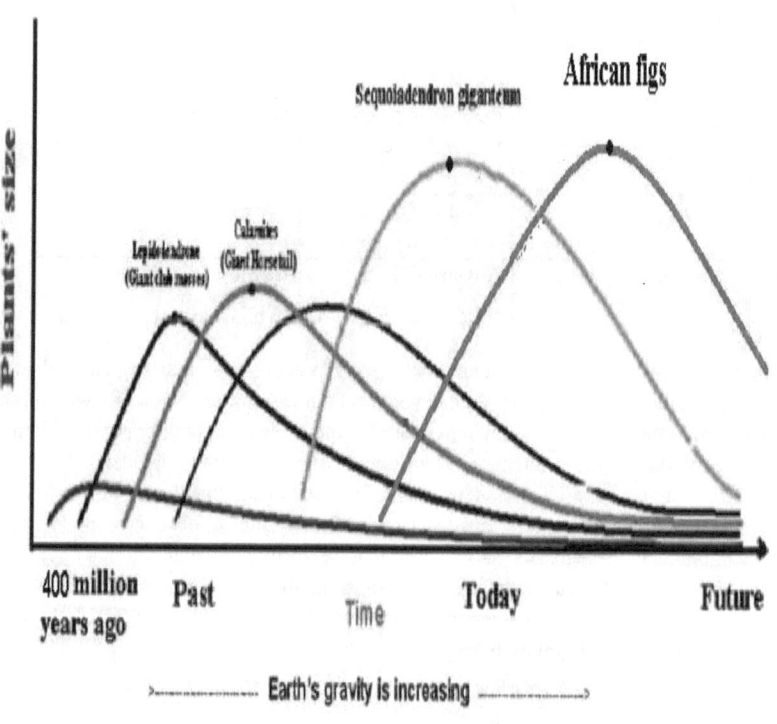

1. The extinction of dinosaurs and the gravity increase

From the earliest days that the big bones of past animals were found, some questions about their decline came up, which continued up to this day. By observing the fossils of giant dinosaurs, about several million years ago, this question comes up that, which factors caused them to decline?

Scientist has made many efforts to answer this question, and they suggested many theories in this case. Theories like the outbreak of diseases, decreasing food and eating of their eggs by other smaller mammals.

Among these theories, there is one theory which is of more acceptance for scientists and attracts the attention of more acceptance for scientists and attracts the attention of most people, this theory which came up by two American geologists from California's university (Berkeley), Luis Alvarez and his son Walter at recent years of seventh decade, is about the collision of meteorite with earth. This theory says that: A rock with 10-15km diameter and about one thousand billion ton weight, with 150 thousand km/h entered earth's atmosphere and friction of this giant rock on the air caused the sublimation of its outer layers. Four or five seconds later the big core of this big meteorite collided the surface of sea and made a big wave with height of 1km. This wave was distributed on the oceans and cover seashores. Collision of meteorite sent a mixture of steam and dust and stones on the earth. All of these stems, smokes and dusts made a big curtain on earth and for several months or perhaps years the earth was in an absolute dark. These events occurred about 65 million years ago and declined dinosaurs.

If the decline of animals was related to a special period, it was possible to rely on this theory, but as we knew, animals at every time, have lost some species. Perhaps, at one time, these declines were very much, but in any case, declines were existed at every periods of time. If we accept that the collision of a giant meteor it with earth and formation of dust and smoke in 65 million years ago caused the decline of dinosaurs, how we can explain of Endocras with 4.5m diameter, about 500 million years ago. At that time did a giant meteorite collide with earth and declined them? Did another meteorite collide with earth 330 million years ago and declined Eogyrinus? Did the collision of meteorite, about 250 million years ago caused big insects to decline? In this case how we can explain the decline of Lepidodendrales plant, Calamitales and giant tree ferns. Did they was destroyed when a big meteorite collide the earth 300 million years ago. The decline of Macropustion and big Diprotodon in Australia, and decline of the awful bird, Phorohacos, which had a head as big as a horse's and length of 3/6m and big beak of 60cm, about 40 million years ago was because of the collision of meteorite? Did the decline of giant Dinornis maximus and Aepyornis maximus in these years was because of the collision of the big meteorite?

All of these shows that, such a theory can not explain declines animals during the long years. But there is another question, which shows the voidance of this theory! The question is that: assume that the big meteorite collide the earth 65 million years ago, and declined the big dinosaurs, why the remained reptiles couldn't get bigger. The earliest reptiles, which were amphibious, were very small, but they could get bigger and made the dinosaurs. With the collision of didn't meteorite, dinosaurs was destroyed, but some of small reptiles survived. Why they make bigger animals!

Then, it becomes clear that, the factor, which caused them to decline, is still remained and doesn't allow big reptiles to get bigger.

So, the collision of the big meteorite with earth and the formation of dust and smoke in atmosphere can not be supposed as the factor of dinosaurs' decline, because, that collision occurred at that time and the effect of dust and smoke have been disappeared and there was no later effect on animals.

Although the theory of the collision of the big meteorite and the formation of dust and smoke can not explain of the animals, but, at the same time, scientists hadn't made a mistake by suggesting this theory, in fact they are close to reality. Because in fact, decline of animals is some how related to meteorites, but not in a way that Luis Alvarez and his son explain.

The theory of the increasing of the gravity says that the fall of meteorites in long periods of time caused the gradual increase of earth, and by increasing of the gravity the ability of blood system of these animals decreased gradually. So those animals which had weaker blood system become smaller or declined. For this reason, after decline of dinosaurs, other reptiles couldn't get bigger, but small mammals which get stronger blood system were able to get their bodies bigger and made the big mammals of 20 million years ago. So, we see that, the scientists are not wrong, and they are right in thinking meteorites as an important factor in this phenomenon. But in the way of explaining and the quality of the extension of the meteorites, they are wrong. Scientists are looking for a big meteorite, which have been able to from such great dust and smoke on earth. But as we know it is not necessary to find a big meteorite, and small meteorites can also increase the gravity provided that their number is large.

People, who accept the theory of the collision of a big meteorite with earth, think that by revealing the existing genes in the remained eggs of dinosaurs they can rebuild them, or by using the sperms of frozen Mammoth they can rebuild them. They think that the decline factor, which was the big meteorite, collided at the time and everything is finished, and they can form new giant animals. These scientist are unaware of the earth which they put their foots on it, and they don't as far the gravity of earth is like the gravity this time, it is impossible to revival such giant animals. Perhaps this is possible in Moon, the gravity of which is 1/6 of that of the earth.

2. The relationship between the blood system and air pressure

we knew that there is a direct relation between the strength of blood system and the largeness of animals' bodies. In away that, the greater the efficacy of blood system and the strength of the heart of animal, the bigger the size of its body. Because, by increase of the strength of blood system of the animal, it can over come the gravity easily and makes the blood to go higher and make its body larger.

But, there is a question. How far the increasing of the increase of heart and as a result the increase of its body's size, can be continued? If an animal has a blood system with good efficacy and a strong heart, to what extent can get longer and enhance its body's size?

Assume that we have a water pump, and we want to draw water from a well. The question is that: How much is the maximum depth, that we can draw water by this pump from the well?

According to Aristotle, it was possible; it was possible to draw the water by the pump from every depth. But the miners, who intended to send the out of the mine, noticed that, they couldn't draw the water more than ten meters depth, even if they pump very hard. At the last years of his life Galileo became interested to this subject, but he couldn't achieve a result, except for this result that the nature hates vacuity, to some extent. He thought that if he uses heavier liquid, wills this amount decrease or not? But before practicing this, he died. His student Torricelli examined this experience in 1644. They conducted this examination with mercury, and they noticed that, they could lift it up to 76cm only in vacuity. They justified this, with air pressure and said that, the air around us have a pressure on the all the surfaces of the earth, the value of which is steady on the Earth, and this pressure causes the coming up of the water from the well through pumping, and because its strength is a clear amount, with what intensity we pump, it is not possible to draw the water more than ten meters (of course now, in deep wells float pumps are being used at the button of the wells and these pumps do not suck the water up- wards, but they push the water from down-ward).

If we suppose the heart of animals like a water pump, we can understand the importance of the air pressure around us. The heart has to pump the blood from lower parts of body. The air interference in this action, and when the heart dilates, it makes a vacuity in itself, and the air pressure, sends the blood toward the heart, by forcing on the surface of the body. As we knew, the force of the air pressure is a certain quantity, and the heart cannot lift up the blood from a certain amount, as the water pump was not able to lift up the water more than 10 meters.

The blood is heavier than the water, and its density is 1.6, that is, so ever strong is the heart, it is able to settle 6m above the ground. So we that the maximum size of animal's body is interrelated with the amount of air pressure. Blood system has more efficiency in less gravity and high air pressure. So ever the strength of blood system is weaker; the animal is less able to overcome the gravity and its size become smaller. Also, if the animal has weak blood system, it has to live in a place that the air pressure is high. So the birds that have strong blood system and advanced respiratory system can fly in high heights. Like Indian goose which flies at 9000m of sea surface where the air pressure in one third of the air pressure on the surface of sea. On earth, big mammals with strong blood system are able to live in most of earth and, they are tall and have large bodies. Reptiles, which have weaker hearts, can only lie on the ground. When amphibious are newborn and have a weak 2-chambered hearts, live in water, and when they obtain a strong and 3-chambered heart they can live on earth but with small bodies. Fish can live in water with 2-chambered heart, and it lies in the water.

The water is 800 times denser than the water and the pressure inside the water are by far more than that of the air. In different depths of water, the pressure is different, and for every 10m depth, one atmosphere is added to the pressure of the water. In deep areas of water, the pressure is very high, so we conclude that, the animals, which have weaker blood system, should be survived in depths of water. So, Coelacanth which is a primary fish, now it is able to live in very deep seas and if we bring it to less deep waters it will die within a short time.

Also the animals which have weaker blood system can be very large, only in very deep water. Like the largest crabs that live in a large deep sea. Because in it place there is so much pressure.

1

The **height/Length** of the animals depends on the following factors:

1-Air pressure

2-Earth surface gravity
3-Density of Blood
4-Strength of the circulatory system
5- Body angle relative to the ground
6- Amount of organic phosphates (like DPG) existing in blood

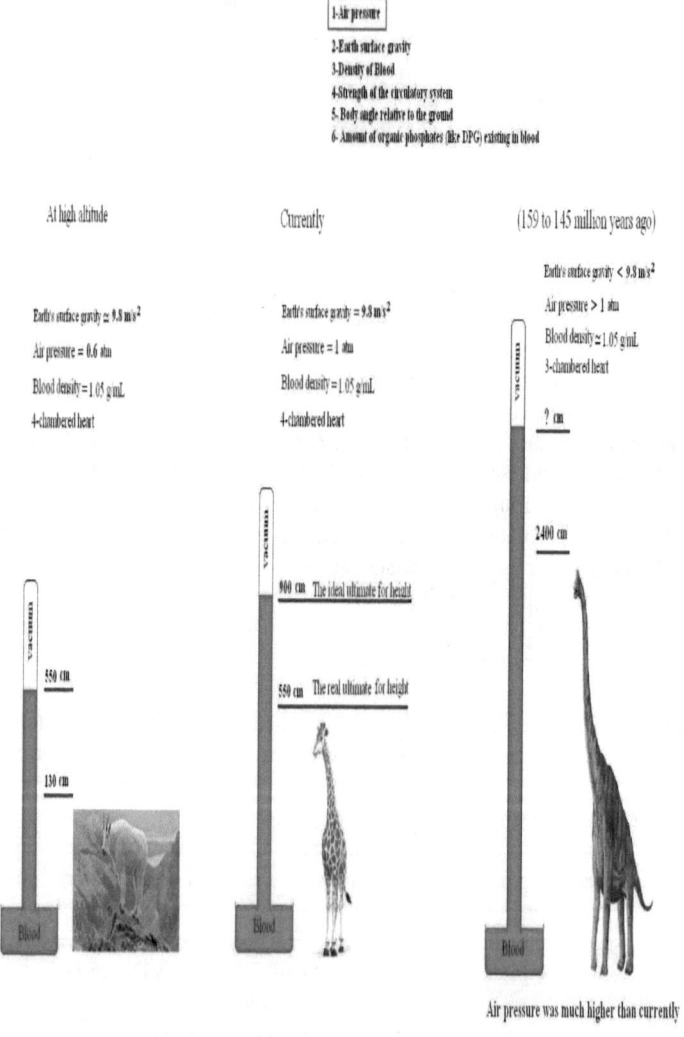

At high altitude

Earth's surface gravity $\simeq 9.8\,\text{m/s}^2$

Air pressure = 0.6 atm

Blood density = 1.05 g/mL

4-chambered heart

550 cm

130 cm

Blood

Currently

Earth's surface gravity = $9.8\,\text{m/s}^2$

Air pressure = 1 atm

Blood density = 1.05 g/mL

4-chambered heart

900 cm   The ideal ultimate for height

550 cm   The real ultimate for height

Blood

(159 to 145 million years ago)

Earth's surface gravity $< 9.8\,\text{m/s}^2$

Air pressure > 1 atm

Blood density $\simeq 1.05$ g/mL

3-chambered heart

? cm

2400 cm

Blood

Air pressure was much higher than currently

Ramin Amirmardfar
6 desember 2014
Rewute 15 December 2015

3

The length of the animals depends on the following factors:

1-Air pressure

2-Earth surface gravity

3-Density of Blood

4-Strength of the circulatory system

5- Body angle relative to the ground

6- Amount of organic phosphates (like DPG) existing in blood

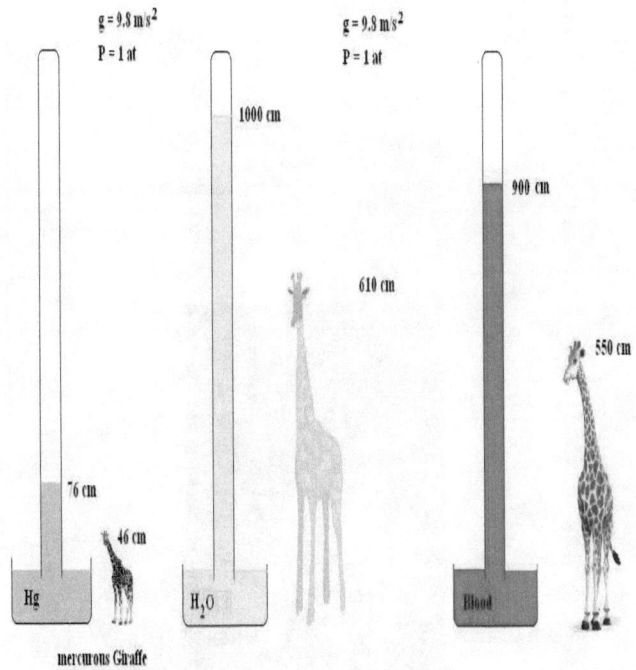

mercurous Giraffe

Ramin Amirmardfar
6 desember 2014

With going down in depth of sea there is a lot of pressure

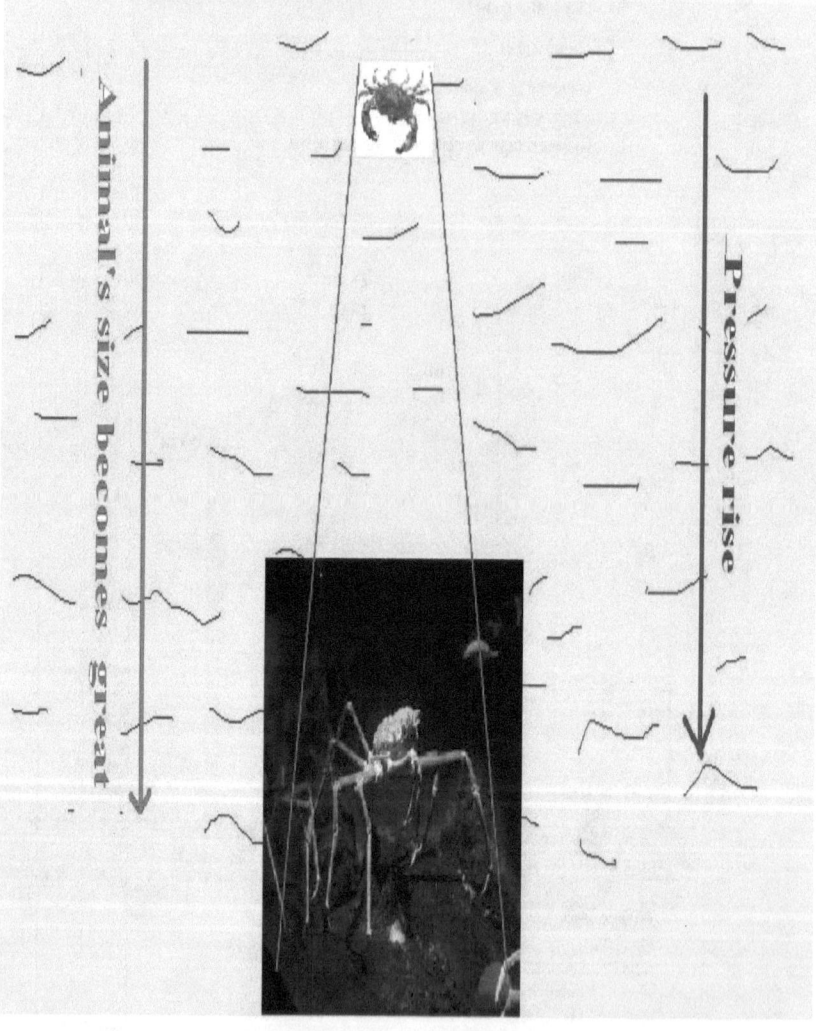

Ramin Amirmardfar

Pressure is lower at high altitudes

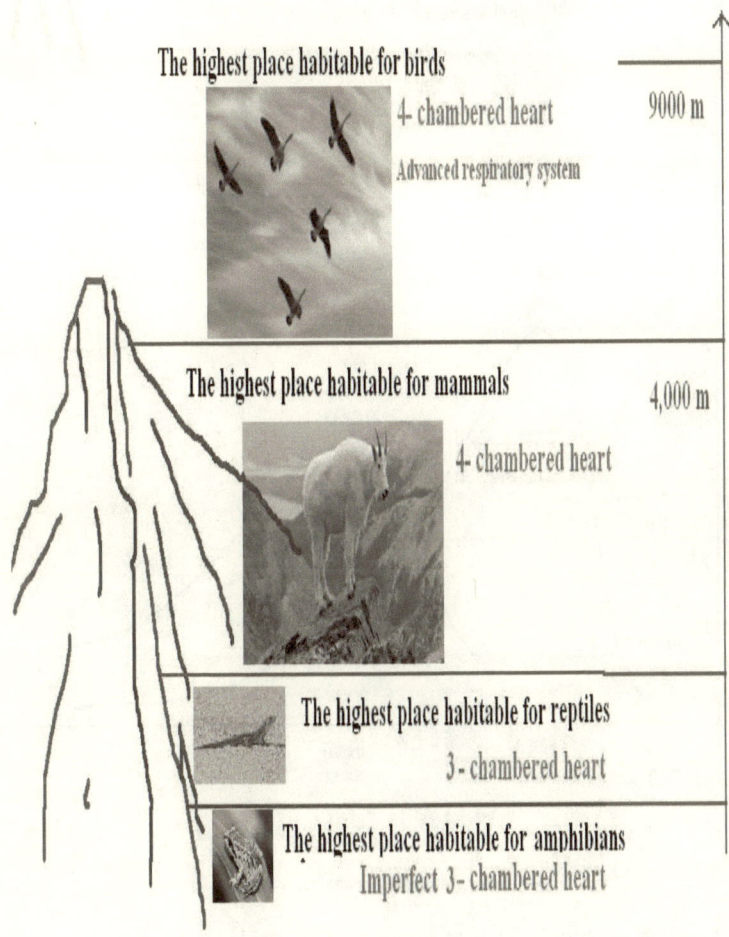

The highest place habitable for birds

4 - chambered heart

Advanced respiratory system

9000 m

The highest place habitable for mammals

4 - chambered heart

4,000 m

The highest place habitable for reptiles

3 - chambered heart

The highest place habitable for amphibians

Imperfect 3 - chambered heart

Ramin Amirmardafr

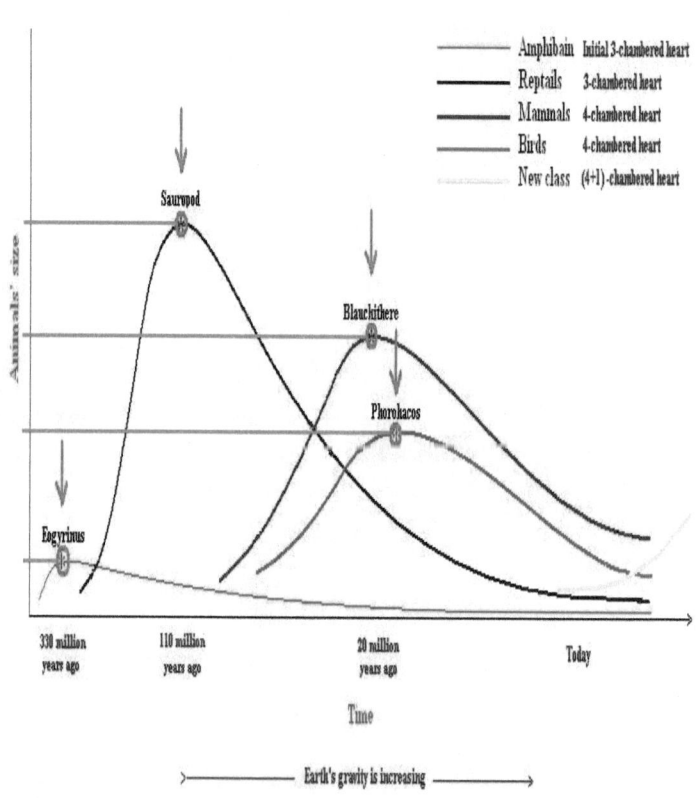

The graph of land animals' size during the time

Graphic by Ramin Amir mardfar

(1995)

The above graph shows that:The amount of air pressure, in the life time of the Sauropod, was higher than all other times.

3. Why did the initial mammals have small bodies?

The first mammals which appeared on earth had the body size of today's small mice. Then by the passage of time their bodies become larger. The well-know example is the species of horse which had small bodies at the time of their emergence, and than they became larger. The horses began from Eohippus which had 28cm in height. Then they gradually became larger and produced Mesohippus, Merychippus, and finally today's horses. The first camels also were very small and were 50cm height, then their bodies became larger (Oxydactylus) and then they converted to Gigantecamelus fricki which had 4m height. After this extremity they began to become smaller until they reached to the height of today's camels which have 2.3m height. the elephants also developed from previous small animals. At first, the elephants were small and they had no trunk and they had rather big front teeth. Then they became larger and larger until they converted to Mammoth and Elephas imperator with 4.1m height. After this extremity they began to become smaller, too. They became a little smaller (Matodon) and finally they reached to the size of today's elephants, that is, with 2m height. Rhinoceroses also developed from previous small animals, and then they gradually became, larger, until they produced larger species of Baluchithere. Baluchithere has been the largest mammal on the earth up to now. But later they made their bodies smaller, too, until they reached to the size of today's rhinoceroses. Deer, like other animals began from small bodies and then reached to the size of Civatherium which was to times bigger than buffalo than its head was as large as elephant's head with a pair of antler on it. They gradually became smaller until they reached to the size of today's deer's. Carnivorous animals also began from small bodies then they became larger until they reach to their extremity, like Maicrodous was similar to tiger but very larger than it, with fangs which were 14cm in length. After this they became smaller and reached to the size of today's carnivorous animals. These examples do not include the mammals only, if we have a look to the genealogy of any animals, we can see that it begins from a small animal and then it gets larger and reaches to its extremity and then, it begins to get smaller again, until it reaches to its current size. The class of reptiles do not need to be explained and all of us are aware of their emergence of first small reptiles from amphibians and their getting larger to dinosaurs and then their gradual getting smaller up to now. The class of birds also began from small bodies. Archaeopteryx, the first bird, which was as small as a pigeon, developed from reptiles. Then they became larger and larger until they produced species as large as Moa (Dinornis maximus) with 4m height and elephant, like bird (Aepyornis maximus) with 3.5m height and 500kg weight, they were living up to several hundred years ago, the first one in New Zealand and the second one in Madagascar. But these birds also made their bodies smaller until they reached to the size of today's ostriches, which had only 2.6m height and 137kg weight. Previously, we knew that by the passage of time, the gravity increases gradually from past up to now, and this causes large animals make their bodies smaller, because of not having enough strength to transmit the blood to high heights in high gravity. When these animals completed their blood circulatory system, they could get their lost strength again and they were able to overcome the high gravity and made their body larger. Large amphibians like Eogyrinuses which had 4.5m height, made their body smaller, because of the increase of gravity. Later some of these shrunken amphibians could complete their circulatory system and became able to complete their circulatory system and convert to first reptiles. These first small reptiles were able to overcome the increase gravity of that days and made their bodies larger by the help of their strong circulatory system, so that they were able to produce such giant dinosaurs. But the increase of gravity continued without interruption and it increased inasmuch as their circulatory system had not enough strength to overcome gravity and they had to make their bodies smaller. Birds and mammals developed from small reptiles an by the help of their strong circulatory system they could make their bodies larger, but by the more increase of gravity again, they had to make their bodies smaller. Up to here, if we accept the increase of gravity, we don't have more problems. But, here there is a question, when mammals to be produced, there were large reptiles, why they did not evolve and why they did not produce the mammals, and why was this evolution among small reptiles. If the mammals were produced from large reptiles of that time, they did not have to make their bodies larger again and waste the time. Also this is true about birds, why the Archaeopteryx did not develop from a large reptile, and why did an small reptile convert to the first bird and why the birds had to waste more time to make their bodies larger, and also we can ask that why didn't the first reptiles develop from amphibians with 4.5m length and why did the emergence of reptiles begging from small amphibians? Suppose that there is an escalator which moves downwards and everything which is put on it is being taken down. Each step has 1cm height that is the first step is 1cm higher than earth, the second step is 2cm higher than earth, the third step is 3cm higher than earth etc. And beside the steps there is a wall on which the height of each step up to earth is written, that is, beside the tenth step number ten is written, which shows the height of 10cm. When the escalator moves downwards, the steps comes down but the numbers are fixed, that is when the tenth step comes down, it is settled in the place of the ninth step, and its height decries from 10cm to 9cm. Now, suppose that different animals stand on the steps of escalator, but in a certain order, in a way that, the number on the wall indicate the height of the animal which stands beside the wall. That is, the mouse with 4cm height stands on the forth step, the cat with 20cm height, stands on twentieth step the rhinoceros with 1m height on hundredth step and elephant with 2m height on two-hundredth step and also giraffe with 4m height on four-hundredth step and the rest of the animals like them. The downward motion of the escalator is like the increase of gravity, which tries to make the body of the animals smaller, draw them downwards and put their step beside the minor number. For example, the mammoth which stood on 4 to 5m step million years ago, has moved downwards gradually by the effect of the escalator's downward movement, and now the elephant which is from its off springs stands on the step with 2m height. Moa (Dinornis maximus) which was standing on four- hundredth step hundreds of years ago, by the passage of time by the effect of the movement of the escalator came downwards and now the number of the wall, beside it shows the number of 260, that is the place of today's ostrich. The ancestors of tiger were on upper steps in the past, but gradually they were brought down by escalator, and now the tiger stands on lower steps, Gigantecamelus fricki stood on 400th step in the past, but now the escalator has brought it to 250 that is the place of today's camels with 2.5 heights. Each moment the escalator moves downwards, that is the gravity lead them to small bodies, but why the escalator which moves dawn wards for million years, couldn't bring all of the animals to the lower step, and after all this time, that the escalator move downwards, there are some animals which stand on higher steps? This is for the reason that, the animals are able to go up the steps. Although the escalator moves downwards, some of the animals jump from one step to higher step and by doing this, they can prevent their more going downwards, but this jumping doesn't all the time and at completely accidental times. This jumping is possible by gaining a stronger circulatory system, that is, any animal which gains stronger heart, has the permission of jumping to upper step, because it is in this case that it can overcome the gravity more, and make its body larger. An small reptile which stood on first steps 200 million years ago, by completion of its circulatory system and by converting to a mammal, it could be able to go up the steps and go up one by one till it reached to upper steps. That is, it generate Baluchitheres, Mammoth, larger camels, horses and other large mammals. We know that the evolution of the animals occurs by gene mutations and natural selection. The evolution of the circulatory system, too, follows this, and also we know that gene mutations occurs accidentally and out of thousands of mutations, one of them may be good for the animal and be selected to clarify this subject we return to escalator again. An animal is standing on a step and wants to gain the power to jump to the upper step. Suppose that, if the animal eats some spinach like $\diamondsuit$, it can be strong and go to upper steps. But this spinach is in cans with closed door and the doors have a lock which is ciphered and its cipher consisted of three numbers and it can be found only by changing the numbers accidentally. Each animal wants to find the cipher of the lock very soon, open the door and eat the spinach, so that it can jump to upper steps unless, the escalator will draw it downwards. At the foot of the steps there is a deep pool and the animals don't want to fall into it and drown, so they should change the ciphered numbers very soon and find the cipher of the lock and open it. Some of the animals find the cipher and can go to upper steps, and at upper step, too, they don't stop their work and attempt to open the other lock, because the escalator don't stop its work and tries to pull them downwards. This competition between animals and the escalator continues. Or in other words, the competition between animals and the increase of gravity continues. But our question has not be answered yet, why finding the cipher always occurs between small animals? If finding the ciphered is accidental, so, the large and small animals should have the equal chance in finding it. But we see that is not so in proactive and small animals have move chance than large animals in finding the cipher and going up the steps, why? That is, if they had equal chance, both large and small reptiles, should be able to find the cipher of generating a stronger heart and convert to a mammal, but we see that, large reptiles couldn't do this and only small reptiles could discover the cipher and convert to mammal and could go up the steps. Here, we see that there is discrimination between the upper and lower steps of the escalator. Every animal which is on lower steps, have more chance to discover the cipher and opens the doors of the cans sooner than the animal which is on upper steps. Because the cipher of the cans is the same in all of the steps, this is justified by one procedure, and that is the number of participants which are in each group, that is the number of the animals in each species. Suppose that two animals of each kind are placed on each step, that is both of them begin to find the cipher of lock it is clear that, this group which has two members, has double chance to find the cipher of the lock. They find the cipher of the lock, sooner than the other one-member groups and find the cipher of the lock sooner that the other groups and go to upper steps. The group which has more members has more chance to find the ciphers very soon and go to upper steps. We consider the numbers of the cipher like the genes of the animals, their chance occurs by reproduction and each offspring represents a chance of number. In each reproduction the mutation occurs and the gene of the children has a little difference with their parent's gene. Out of thousands of offspring, only one of them has the cipher of opening the lock, that is, one of them has the gene of generating a strong circulatory system and only it is able to go to upper step, that is, the natural selection selects this gene which contains the correct cipher, if the number of the reproduction in one species is low, the number of the changed ciphers is less, but if a lot of animals reproduce together the number of the offspring is more and the chance that one of them carries the correct cipher is high. Now, let's have look to the nature, which animals have high ability of more reproduction and the number of which is high? Let's recall the mouse, that can bear several offspring in a short time and these offspring's became mature after a short time and can bear offspring's again. Let's recall a cat that can bear several offspring each time but only once a year. Then it takes one year until they become mature and begin reproduction. So the rate of reproduction of cat is less than the mouse, so it has less chance to find the ciphers of the locks. Sheep bears only one lamb in a year, and it takes one year until it became mature, so the chance of the sheep is less than the cat. Let's recall the elephant. The elephant bears an offspring, in several year and this offspring becomes mature after several year and then it bears an offspring and during the period that an elephant bears and offspring, the mouse reproduces hundreds of mice like itself and all of these offspring try to find the cipher of the lock together. The number of the mice became more and more and they change the cipher of the locks rapidly, that is, they generate new genes. It is clear that, the chance of appearing a desirable gene among the numbers offspring's is high and possibly one of the mice has the desirable gene and is able to go to upper step. During this time the elephant has tested only one cipher of the lock and it does this very slowly. The number of different genes which is generated by a population of mice in one year needs thousands of years to be generated by the elephants. Because the escalator don't differentiate in downing the steps and don't notice the hugeness of the elephant and pulls them down with equal speed. It is clear that the mice can compensate the downward movement by going up the steps, but the elephant goes down and have no haste to find the cipher of the lock and eat the spinach. Mammoths were 400th step 2-million year ago, but because of low reproduction they couldn't open the lock to eat the spinach and get stronger and also they couldn't go up the steps so they came down along with escalator and now their children, that are elephants, are on 200th steps. 200 million years ago the primitive mammals were very small with high strength of reproduction, they could find the ciphers rapidly and go up the steps and generate large mammals. Here there is an interesting thing, this system is automatic. In nature there is a converse relationship between the largeness of the body of the animal and the power of its reproduction, and all of us know that, the larger the animal, the more the period of its reproduction and its maturity and the number of offspring is, so the number of its population is less. Instead, the smaller the body, the shorter the period of reproduction is, as a result the population of these species is high. We can say that by decreasing the size of the body the speed of finding the cipher is high and by increasing the size of the body, the speed of finding the cipher is slow. Now, again consider the escalator, the escalator tries to draw animals downward, instead, the lower the animal comes, the speed of finding the cipher gets more and the animal goes up more, by going up, the speed of finding the cipher declines and the escalator can draw it downward. The increase of gravity draws bodies animals toward small sizes, instead the smaller the body of the animal gets, the more increase the speed of its reproduction, and it can generate different kinds of genotypes and discover the genotype a strong circulatory system and by doing this, it can overcome the gravity more and gets larger. By increasing the body size, the speed of reproduction lowers again and the animal yields to gravity, and this is repeated continuously. So we see that whatever the escalator works, it is not able to pull all the animals down the and although the gravity is increasing for million of years gradually, but instead the animals learn how to overcome it and make their bodies larger. So, the result is the beginning of the next evolved class, is from the animals of the previous small class. Therefore can we say that now is if the next class wants to be generated, the offspring of which today's species. So, the result is that, the commencement of the next more evolved class, is always from the small animals of the initial class. There can we say that now is if the next class wants to be generated, the offspring of which today's species will they be?

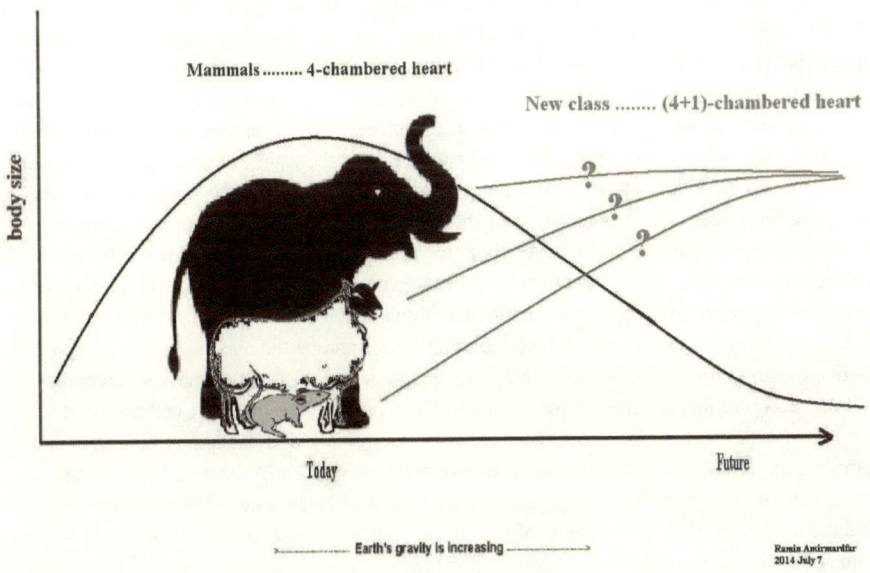

Mammals .... 4-chambered heart

New class .... (4+1)-chambered heart

body size

Today

Future

Earth's gravity is increasing

Ramin Amirmardfar
2014 July 7

4. Why the body size of the mammals are different?

Why the body size of the mammals is different? Previously we knew that there is a direct relationship between the strength of the circulatory system of animals and the largeness of animal and their bodies. The animals which have a weak circulatory system are smaller and which have a complete and strong circulatory system is bigger than the other. So, we see that there are more differences in body size of mammal. The mammals, which have complete and strong circulatory system, must have large bodies, but in fact, it has not been so, and we see small bodies among them! What is the reason? Previously we knew that small animal's whit rapid and more power of propagation, are adjusting themselves with natural changes timely, and continue to survive their generation. In fact to struggle whit the changes of nature, specially, specially the increase of gravity, the animals of each class should remain small enough to make desirable motions and to produce evolutes, whit their rapid propagation. Otherwise, if all of the animals belonged to one class were large, they will not be able to produce desirable animals in appropriate time, so they will be extinct. For example when dinosaurs existed, if all of the reptiles were large, and there were no small reptile, they wouldn't be able to evolutes their circulatory system against the increase of gravity in appropriate time and produce mammals, and they would be extinct by the increase of gravity. But the nature, by the use of suitable methods, maintain some species smaller despite their complete and strong circulatory system, so that they could have enough speed of propagation to produce desirable genes. But how does the nature do this? How can prevent the growth of the body of the animal which has strong and complete circulatory system and keep it smaller? Consider a mountainous area on which some villages are located. Some of the villages are located on the foot of the mountain and some of them are located on the heights. A railroad connects these villages to each other and the train moves on this railroad and brings food for these villages. In animals, the residents of the villages are the cells of animals' bodies. The vessels are like rail and the heart is like the engine of the train. As we knew, there are no blood vessels in insect, that is they don't have railroad, so the train is able to go to farther village and brain food for them, as a result, the residents live only in the village is near to the heart. That is, the body size of the insect must be small. In the amphibian and reptiles there is railroad and also train, but the power of the engine of the train is less and can not draw the train to upper parts so the train to upper parts so the villages have to be arranged horizontally and not to be situated in mountainous areas and high places. That is the body of the animal can't be situated vertically and it has to lie down on earth. In mammals which have both the railroad and the train whit powerful engine, the villages are able to be situated even on the top of the mountains and its residents can receive food and other requirements. That is the mammals can make their bodies larger and they can make the themselves as tall as the giraffe. But we see small bodies among the mammals, too. In their bodies there are both railroad and the train with powerful engine, so why do they have small bodies? The railroad acts like the vessels and the engine of the train acts like the heart and the heart and the train itself acts like the blood. But there are red cells in the blood containing hemoglobin which carry the oxygen. It is like this that there are boxes on the wagons of the train in which the food is being carried and when the train passes through one of villages some of these boxes are opened and the food is given to the residents of the villages. Then the train arrives at the other village and some of the boxes are opened there and the food fall in the village and the villager use them, so as the train passes through the villages, some pf the boxes are opened and the residents receive their required food. At last, when the train arrives at the last village, the boxes are opened, and the train become empty, and returns to the foot of the mountain, loads there, and repeats the operation. In the blood of the animals the red cells do an action like the action of the boxes of the train. At the time of departure, they become full of oxygen and when passing through the tissues empty their oxygen by turn and when all of them empted their oxygen they return to their initial position, and repest the operation again. But when the train passes through a village. How many of the boxes should be opened? The train arrives at the village, all of the boxes are opened there, and the foods fall there. The train moves to the other village, but there is no food to give to its residents. Although, railroad continues up to the upper villages and the engine has the power to go up, but there is no food to give to their residents, so the train returns. So the residents of the upper village die because of the lack of the lake of the food, and the village disappears. That is the body of the animal remains small. Also in the body of the animals, if all of the red cell of the blood release their oxygen in near distance such an action will occur and only the cells and tissues which are near to the heart will survive. That is the animal has to remain small and has not to get larger. In the body of a mouse such event occurs, and all of the oxygen separate from the hemoglobin whit in near distance red cells is emptied. In other animals which are larger, the oxygen attaches to the hemoglobin intensively and is emptied some farther. So according to the largeness of the body of the mammals, the force of attachment of the oxygen to hemoglobin becomes stronger, and it can course a long distance in the body of the animal. But, which factor controls the degree of the attachment of the oxygen to hemoglobin and its early or late release? Organic phosphates like DPG which exist in the combination with hemoglobin control this operation. The more the amount of combined DPG with hemoglobin is, the less the tendency to intake oxygen will be, and vice versa, the less the amount of combined DPG with hemoglobin is , the more the tendency to intake oxygen will be. In the blood of small mammals like mouse, there is a lot of DPG. So, the tendency of their hemoglobin to combine with oxygen is less. While in the body of an elephant, the hemoglobin of the blood has more tendencies to combine with oxygen. Oxygen pressure at which, the hemoglobin gives, most of its oxygen to tissues, is 45mmHg for a mouse, 42mmHa for a rat which is larger than mouse, 38mmHg for a cat, 35mmHg for a fax, 30mmHg for a sheep, 25mmHg for a horse and 22.5mmHg for an elephant which is larger than all. That is in elephant's body the boxes on the train are opened gradually when the train passes through the villages, and the food remains until the train reaches at the highest village and the residents of that village receive food. In sheep's body the boxes are opened in large number at the villages, and the food finishes at the middle of the way and the residents of the upper villages do not receive food. In mouse's body all of the boxes are opened at the first village and the train is emptied very soon and it has to return. So, the course of the train is very short. So it becomes clear that, the nature can control the body the body size of the mammals by use of an organic phosphates beside the hemoglobin and keep some of them smaller in order to more probation. Of course, more propagation needs a lot of energy. Each cell of mouse's body consumes a lot of energy at the unit of time in compare with elephant cells. Because the train inside the body of the mouse empties all of its food at one village and it is clear that each resident of the village will receive a lot of food and so they will have enough food to have more activity. But in elephant's body the food is rationed and only a little food is given for each village, so that all the villages can receive food. For this reason each of the residents of elephant's body receives a little food, so they can not have more activity. Therefore the cells of elephant's body have low metabolism in compare with mouse's cells. As a result the growth and propagation in a mouse is by far more than an elephant. The heart of an elephant, which has 3 tons weight, beats 46puls/min and the heart of a mammal which has 70kg weight beats 76pulse/min. The heart of a cat which has 1.8kg beats 240pulse/min, and the heart of a mammal which has 100g weight, beats 800pulse/min. In heart of a mammal which beats 76pulse/min, the contraction of the ventricle lasts about 0.31s and after that, it rests for 0.49s. That is, the ventricle works 9.5h in 24h, and rests for about 14.5h. In heart of a mammal with faster pulsation, the period of rest is short but the numbers of pulsations are more. In a mammal whose heart beats very fast the contraction of the ventricle is 0.024s and the period of rest is 0.036s. So the ventricle works 9.5h a day and the remainder of ther day it rests. As we see the period of work and the period of the rest of the heart during a day, in large and small mammals is almost equal. So, the result is that, the strength of heart in mammals with different size is almost equal. When the body becomes larger the rate of heart becomes low and instead, its strength becomes more. When the body becomes smaller, the rate of the heart becomes faster, but instead its strength becomes less. But eventually, the whole energy and power of the heart remains fixed. This is like the changing of the automobile's gear. In an automobile the power of the engine is fixed. When the automobile moves with heavy gear, speed is low, but the power is high. When it moves with light gear, the speed is high but the power is low. In these two cases the power of the engine does not get more or less but, the gearwheels and levers change the energy to speed and force. Also in the heart of a mammal, such an action occurs. The capabilities of heart in all mammals are almost equal and only the rate of speed and force gets more and less.

5. The amount of organic phosphates (like DPG) existing in hemoglobin of blood is determining factor of mammal's bulk

Mammals are found in different sizes in nature. Most of scientists believe that environmental circumstances and mammals' living type determine their body size. So scientists haven't been looking forward to find a physiological factor for this size difference among mammals. But my research on animals and plants size has specified that there is a direct relationship between animals/plants size and blood/fluid circulatory system.

Mammal species/oxygen pressure in blood

The oxygen molecules enter into the blood from lungs and stick to the red cells. After moving a distance they reach the tissues and get separated from the red cells and then the cell use them. What the oxygen molecules combine with hemoglobin and stick to the red cell? What causes the oxygen molecules be separated from the hemoglobin in the tissues? And the most important is that, what causes the oxygen molecules be separated from the red cells soon or late?

The transfer of oxygen to the blood is done by circulation. So the reason is the oxygen's thickness or pressure. The oxygen's pressure in lungs is more than the lung's capillaries. So the oxygen enters into the blood from the lungs. These oxygen molecules mix with the hemoglobin in the red cells, then move to the heart and after it, move to the tissues. When these molecules reach near the tissues, they get separated from the red cells because of the lowness of oxygen pressure. But this separation is different among the mammals' species. Most of the oxygen molecules get separated from the red cell's hemoglobin in mouse's body in 45 mmHg. But this happens in 22.5 mmHg in elephant's body.

The organic phosphate/the combination of oxygen with the hemoglobin

It means that the oxygen molecules combine with hemoglobin very weakly in mouse's body. Getting a little far from the lungs the oxygen's thickness gets weaker and so they get separated from the hemoglobin and so the red cells can not move for longer distances. But in elephant's body have stronger stickiness; the oxygen's pressure gets lower by getting far from the lungs. But most of oxygen molecules still stick to the hemoglobin's and move to the far tissues. On the other hand, the mouse's tissues can not be far from the lungs and heart because the oxygen can not move a long distance but it is different in whale and elephant's body because they are sure that all their tissues will get oxygen. Now we have question, why do we have this difference in stickiness among the mammals? To say it simply, why the combination of oxygen molecules with hemoglobin is stronger in elephant's blood than the mouse?

The answer to this question relates to the organic phosphate like (DPG) which exists in mammal's blood. The more organic phosphate the less the stickiness between oxygen and hemoglobin, like the hemoglobin of mouse. On the contrary, the less these organic phosphate the more the stickiness power, like the elephants body.[8]

Simply, organic phosphates make the stickiness between the oxygen molecules and hemoglobin weaker so there is an opposite relation between the amount of organic phosphate and the stickiness power. The less the amount of organic phosphates, the larger the size of mammals.

Mouse	....	**45.0** mmHg
Rat	....	**42.0** mmHg
Cat	....	**38.0** mmHg
Fox	....	**35.0** mmHg
Sheep	....	**30.0** mmHg
Horse	....	**25.0** mmHg
Elephant	....	**22.5** mmHg

Fig.1 The mammal species and the oxygen pressure in which the red cell's hemoglobin sends most of it's oxygen to the tissues. In elephant's body have stronger stickiness, the oxygen's pressure gets lower by getting far from the lungs. But most of oxygen molecules still stick to the hemoglobin's and move to the far tissues. But the mouse's tissues can not be far from the lungs and heart because the oxygen can not move a long distance.

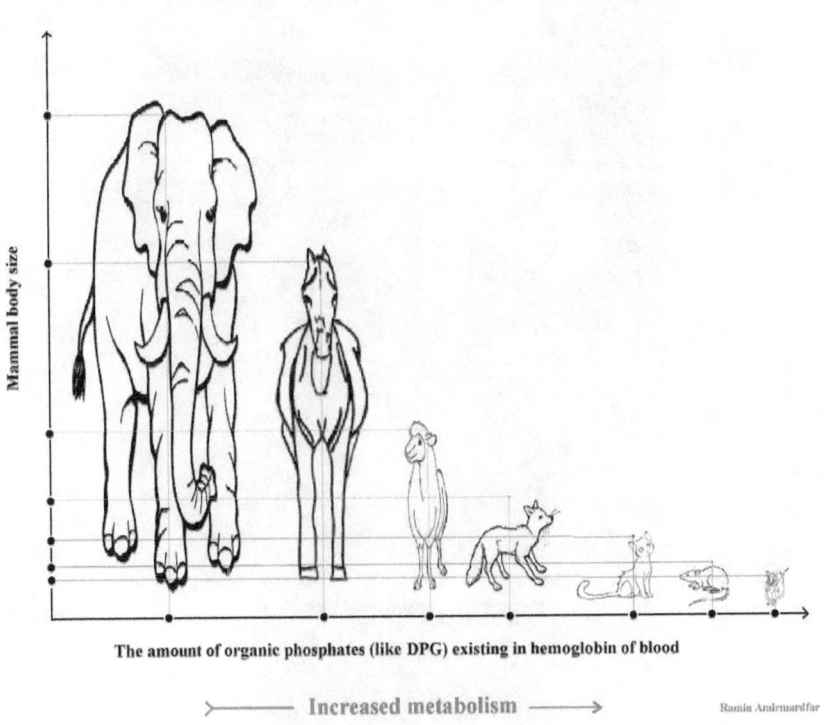

The amount of organic phosphates (like DPG) existing in hemoglobin of blood

⟩ ——— Increased metabolism ———⟩   Ramin Aralemardfar

Fig.2 "Y" axis represents body size of mammals. "X" axis represents the amount of organic phosphates (like DPG) existing in blood. Organic phosphates make the stickiness between the oxygen molecules and hemoglobin weaker so there is an opposite relation between the amount of organic phosphate and the stickiness power. The less the amount of organic phosphates, the larger the size of mammals.

Conclusions

Physiological factor of amount controlling of mammal's bulk is the amount of organic phosphates like (DPG) existing in red blood cells of mammals. Organic phosphates make the stickiness between the oxygen molecules and hemoglobin weaker so there is an opposite relation between the amount of organic phosphate and the stickiness power. The less the amount of organic phosphates, the larger the size of mammals. (Fig. 2)

Scientific application of this science

We can use this science in applied sciences (like husbandry and nurturing domestic animals) and maximize or miniaturize body size (bulk) of domestic animals due to necessities of poulterers. Of course human has done this work previously unconsciously on domestic dogs. Human has created dogs in different sizes artificially without any awareness of mechanism of this action. (Figs. 3a, 3b) (Figs. 4a, 4b)

By the help of this science we will be able to create domestic animals in desired size consciously and open-eyed. (Figs. 5a, 5b) (Figs. 6a, 6b)

(a)

(b)

Fig. 3 Amount of DPG in the blood is low, therefore dog is large; (a) Amount of DPG in the blood is high, therefore dog is small; (b)

(a)   (b)

Fig. 4 Amount of DPG in the blood is low, therefore horse is large; (a) Amount of DPG in the blood is high, therefore horse is small; (b)

(a)   (b)

Fig. 5 Amount of DPG in sheep blood is high, therefore sheep is small; (a) If amount of DPG decrease in sheep blood, therefore sheep will larger; (b)

(a)   **(b)**

Fig. 6 Amount of DPG in cow blood is low, therefore cow is large; (a) If amount of DPG increase in cow blood, therefore cow will smaller; (b)

Here I propose a solution to the preceding two questions.

I use an analogy. Let us consider a village with its in habitants. There is an apple garden near it. A person should feed the inhabitants from this garden. This person has only one tray picks up the apples and puts them in the tray, and then he goes to the village. On the way, some of the apples slip from the tray. Because the village is near, he could carry some of them to the village. But if the village was far away, all the apples would slip from the tray and the inhabitants would die without apples. So this person should find a way to carry the apples for a long distance. We compare this example with the mammals' body.

The villages are like the body's tissues, the in habitants are the cells, the garden is the lungs, the apples are the oxygen molecules, and the way is the blood vessel. That person is the blood and the tray is the red cells. Here, the thing that causes the oxygen molecules stick on the red cells is our point. Now we study how oxygen molecules stick on the red cell. There is hemoglobin in red cells. The oxygen molecules are joined with the hemoglobin and this causes that the oxygen molecules stick to the red cells. With little considerations, we find that the oxygen molecules do not stick to the red cells with a similar strength in the body of all the mammals. This stickiness is much stronger in some species and is weaker in some other. This stickiness is related to the size of the mammal. The more this stickiness, the more the oxygen molecules can be carried in the blood. This will cause the body to be larger. On the contrary, if the stickiness is weaker, the oxygen molecules would move less in the blood and so the body has to be small.

Look at the following information:

The mammal species and the oxygen pressure in which the red cell's hemoglobin sends most of it's oxygen to the tissues.

Mouse ....45 mmHg

Rat ....42 mmHg

Cat ....38 mmHg

Fox ....35 mmHg

Sheep ....30 mmHg

Man ....28 mmHg

Horse ....25 mmHg

Elephant ....22.5 mmHg

The oxygen molecules enter into the blood from lungs and stick to the red cells. After moving a distance they reach the tissues and get separated from the red cells and then the cell use them. What the oxygen molecules mix with hemoglobin and stick to the red cell? What causes the oxygen molecules be separated from the hemoglobin in the tissues? And the most important is that, what causes the oxygen molecules be separated from the red cells soon or late?

The transfer of oxygen to the blood is done by circulation. So the reason is the oxygen's thickness or pressure. The oxygen's pressure in lungs is more than the lung's capillaries. So the oxygen enters into the blood from the lungs. These oxygen molecules mix with the hemoglobin in the red cells, then move to the heart and after it, move to the tissues. When these molecules reach near the tissues, they get separated from the red cells because of the lowness of oxygen pressure. But this separation is different among the mammals' species. As it is shown in the table above, most of the oxygen molecules get separated from the red cell's hemoglobin in mouse's body in 45 mmHg. But this happens in 22.5 mmHg in elephant's body. It means that the oxygen molecules mix with hemoglobin very weakly in mouse's body. Getting a little far from the lungs the oxygen's thickness gets weaker and so they get separated from the hemoglobin and so the red cells can not move for longer distances.

But in elephant's body have stronger stickiness; the oxygen's pressure gets lower by getting far from the lungs. But most of oxygen molecules still stick to the hemoglobin's and move to the far tissues. On the other hand, the mouse's tissues can not be far from the lungs and heart because the oxygen can move a long distance but it is different in whale and elephant's body because they are sure that all their tissues will get oxygen. Now we have question, why do we have this difference in stickiness among the mammals? To say it simply, why the mixture of oxygen molecules with hemoglobin is stronger in elephant's blood than the mouse?

The answer to this question relates to the organic phosphate like (DPG) which exists in mammal's blood. The more organic phosphate the less the stickiness between oxygen and hemoglobin, like the hemoglobin of mouse. On the contrary, the less these organic phosphate the more the stickiness power, like the elephants body.

Simply, organic phosphates make the stickiness between the oxygen molecules and hemoglobin weaker so there is an opposite relation between the amount of organic phosphate and the stickiness power. The less the amount of organic phosphates, the larger the size of mammals.

Comparison of the cars with animals

Car	Animals
Size	Body size
Engine power	Heart power
The number of engine cylinders	The number of heart chambered
Movement speed	Metabolic rate
Reducer gears	Blood's organic phosphates (DPG)
Efficiency range	Living range

Ramin Amirmardfar

15 January 2015

"Reductions speed-increase size" by using gear on the 4-cylinder car

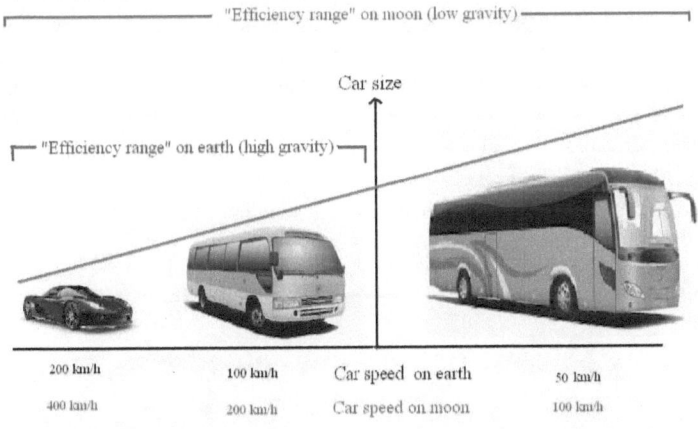

"Efficiency range" on moon (low gravity)

Car size

"Efficiency range" on earth (high gravity)

| 200 km/h | 100 km/h | Car speed on earth | 50 km/h |
| 400 km/h | 200 km/h | Car speed on moon | 100 km/h |

Ramin Amirmardfar
15 January 2015

"Reductions metabolism rate-increase body size" by using blood organic phosphates (DPG)

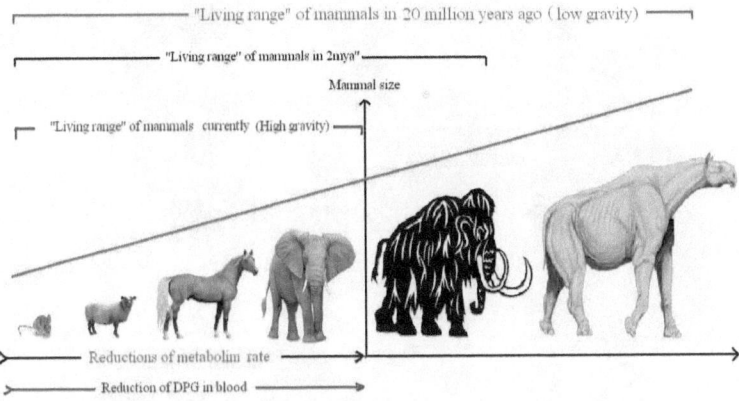

"Living range" of mammals in 20 million years ago (low gravity)

"Living range" of mammals in 2mya"

Mammal size

"Living range" of mammals currently (High gravity)

Reductions of metabolim rate

Reduction of DPG in blood

Ramin Amirmardfar
15 January 2015

6. New heart of vertebrates in the Future

We understood that the gravity has been increasing since the past and animals in order to better overcome this high gravity have been forced to build themselves stronger hearts: one-chambered, two-chambered, three-chambered and finally four-chambered hearts. As we know the largest and the most successful animal at present are those who have 4-chambered hearts (e.g. Mammals and Birds). We also understood that the increase of gravity is now continuing. It is clear that the 4-chambered heart, too, in near future, will lose its power of overcoming the gravity and the animals will have to build stronger hearts. That is the question: Are animals thinking of building stronger hearts? Have they taken measures about this? Which method they will use to strengthen their blood-circulation system?

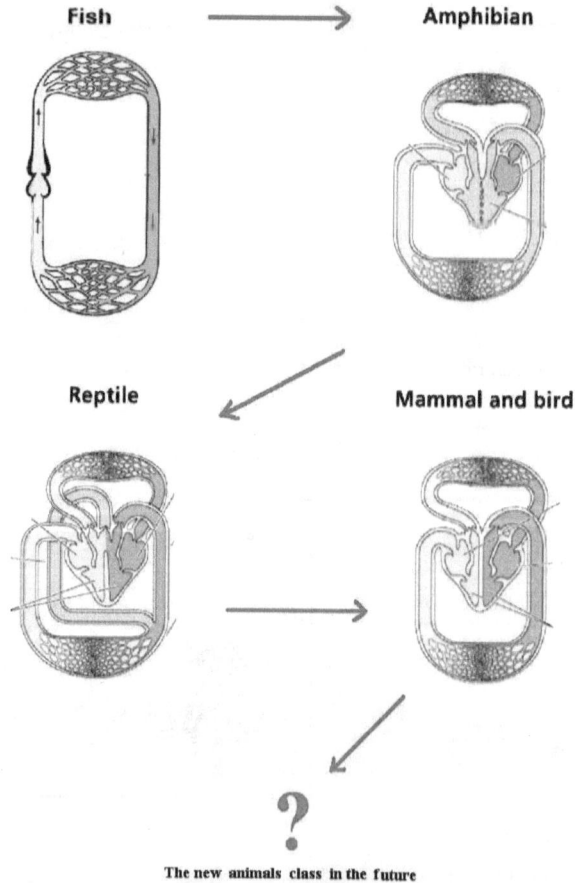

The new animals class in the future
will have (4+1)-chambered heart

Ramin Amirmardfar

Consider a person who has stood motionlessly in a place for some time. If he continues his motion less standing, because of not receiving enough blood in his brain, he will swoon and fall. Why dose this occur? Why does a person swoon when he stands motionless in a place? Why this is not the case when he walks and his brain receives enough blood? In both of these case, the gravity affects in the same way, the heart in both of them has the same power, in the first case the brain receives enough blood but in second it does not. Why? Because in the second case (walking) some small hearts in other parts of the body help the blood- circulation system and provide enough blood to the brain, but in the first case (motionless standing) these small hearts don't work, so the brain dose not receive enough blood.

Where have these small hearts been located in the body? And how they work? These hearts are parts of the vein which are surrounded by two venous valves. When a person is standing motionless, these hearts do not work but when he starts walking, muscle of foot tightens these

veins and makes the blood run out side. Because there are some valves inside these veins which prevent the blood from running downwards, so the blood is always run upwards. In every contracts and expands of muscle of the foot, this part of the foot vein expands and contracts like a small heart and carries the blood upwards. Therefore, if is clear that this person's heart cannot circulate the blood by itself and is not able to overcome the gravity and send enough blood to the brain. If these small hearts do not help the main heart the person will swoon and fall even while walking, which can occur several times a day. The only failure of these small hearts is that they work just when the animal is walking because they receive their power from the muscles of hands or feet and are not able to expand and contract by themselves.

Wasn't it better that these small hearts had independent muscles for themselves and could contract without the help of hands and feet muscles? In this case, even in motionless state, the brain could receive enough blood and the animal would not swoon. If we consider other classes of animals like Mollusca, we would find out that in the body of some of these animals there are such small independent hearts that help blood circulation system. Octopuses have auxiliary heart in front of their gills that effectively help the main heart in blood circulation system. So it is evident that the nature has already used auxiliary hearts. Nature has also provided the requirements of building auxiliary hearts in the body of Mammals. Just a small muscle is needed in order for a small heart to work. Thus we see that the animals are thinking of stronger hearts for themselves and in the future, there will be animals which, as well as their main heart, they will have auxiliary hearts in their feet that help effectively blood-circulation system.

The increase of gravity will cause the bulk of animals be more smaller than before and those who are not able to face gravity, will have to crawl. Elephants, Rhinoceroses and huge Giraffes will not

exist, they will either be extincted or have smaller bulks (There are Elephants with smaller bulks in Africa at present).

68

Some coastal Mammals (like Seals and Walrus) will not be able to come to the shore any more and will become sea living animals (like Whale, Manatee and Dolphins). Mammals in the future will not have the power of standing erect and will have to crawl (like class of Reptiles which were forced to do so a long time before). Large birds (like Vultures) will not be able to fly and will live on the ground (like Ostrich which was forced to do so already). Small birds, too, will hardly fly.

But some animals out of the Mammals will reach those auxiliary hearts. These animals will overcome to the high gravity and larger their bulks and stand erect. They will provide blood to their high altitudes by the help of their powerful blood- circulation system and will rise their head and won't have to crawl.

As consequence this final question arises: among Mammals, which species have the high chance of reaching a powerful blood-circulation system? After the evolution of their blood circulation system, will they be remained Mammals? Or will they found a new class of animals? What other characteristics they will have?

7. New and more efficient respiratory system in the future vertebrates.

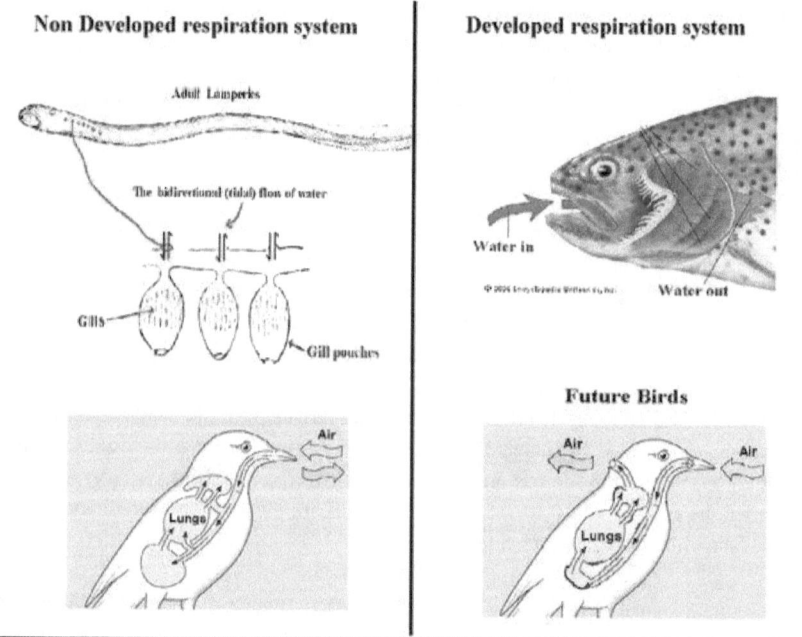

Non Developed respiration system

Developed respiration system

Future Birds

Ramin Amirmardfar
2013

8. The Pace of Evolution and its Relation with Continental Drift and Expanding Earth

Let us begin the matter with a question: What is your opinion about why native people of continents of America and Australia were living in primitive societies before the arrival of civilized people from other continents? Why civilized people had not been appeared in these two continents?

The development of humans is dependent to the development of science and development of science is dependent to scientists of the society. The occurrence of scientists in each society is accidental and follows statistical rules.

Scientists leave their inventions and discoveries to scientists who come later, and the recent generation of them uses this information to find or invent new other things. Thus, science is developing in the society and is getting progresses. The pace of evolution of science in each society is in connection with two things:

1) The percentage of scientists existing in the society;

2) The rate of connection between scientists.

The possibility of appearance of scientists in each society has a direct relation to the population amount of the society. This means that, more populated is the society, more the scientists that exist in that society. That is the reason why the most development has come to existence in great and populated continents like Asia- Europe-Africa.
If a scientist has not access to information of scientists previous to him, he must independently to find or invent all the in- formation from the beginning. In this case, he will not have enough time to find or in- vent new matters and his work leads to a meaningful slump of the pace of scientific evolution in that society. By a glance to the different continents of the Earth we can observe easily that:
Continents like Asia-Europe-Africa which have had a connection each other have had a scientific development and a quicker pace of evolution because of the mutual connection of their scientists and of the exchanging of information. On the contrary, continents like Australia and America (before getting explored) that were isolated and without connections, have not had a noticeable scientific development, and additionally both their pace of evolution was slower and their population were living primitively. The northern parts of Africa that were in connection with continents Asia and Europe, were more developed than the southern parts, which were more far and without connection.
Also now, if we cut the entire connection of the Australian continent with other lands of the world, after a short period, we will be witnesses of a scientific lag of that continent with respect to other continents, because the population of Australia is lesser than the population of the rest of the world. This is why the percentage of appearance of scientists will be less there, and consequently the development of science in Australia will be less than the average one in the world. If there was not any sea-route or land- route between Asia and Europe, the science of making the compass, the paper, the printing, the banknotes and the gunpowder would not reached Europe from Asia, and

Europe would not have developed up to the present extent. And, on the contrary, if new sciences was not arrived to Asia from Europe, now Asia would be in great lag.

I think that this example – an analogy – makes clear to us that into the pace of evolution of societies, it has been involved the rate of connection between them and the amount of population. We use this example as an analogy for the pace of evolution of land animals. The appearance of a proper genotype for development and evolution of land animals – like the appearance of the scientists – is an entirely accidental issue and follows the probabilities rules applied to the population of land animals. More the population, more the percentage of appearance of land animals possessing the desired genotypes. And more the connection of various regions with each other, more the diffusion of desired genotypes.

The life place of land animals is on the surface of Earth's continents, so, if continents are larger and their forms are more circular, the pace of evolution of land animals of that continent will be more quick; indeed, the pace of evolution of the land-animals in Asia, Europe and Africa is much more fast than the American and Australian ones. If continents had not separated from each other, remaining a constant connection among them, the pace of evolution would has been greater and now we would be witnesses a number of new classes.

Here, I mention its suitable example. The modern mammals (placental mammals) on their evolutionary path have passed monotremes and Marsupials. In Asia and Europe marsupials had existed already in preceding epochs. But later, modern mammals have evolved from them. Since these continents are great and they have had intense mutual relations, as con- sequence the pace of evolution of animals in these continents has been faster and can have produced the modern mammals. But the continent of Australia has had slower pace of evolution due to its small surface area and lack of connection with other places, with the result that over there mammals are in primitive stages and have remained marsupials. If modern mammals were not entered in Australia from other continents, for the appearance of modern mammals might be needed millions of years due to its very slow pace of evolution.

By these interpretations, I think it is clear what I mean to you. I mean that the pace of evolution can be a good evidence for the theory of expanding Earth. According to the theory of expanding Earth, all the continents were contiguous in the past and were in connection with each other from every side. On this kind of super continent, the land fauna should have had a very high pace of evolution.

Appearance of the first vertebrates (early amphibians) in land	370 million years ago	
The first appearance of reptiles from amphibians	350 million years ago	150 million years elapsed time for the evolution of reptiles from amphibians
The first appearance of mammals and birds from reptiles	220 & 150 million years ago	150 million years elapsed time for the evolution of mammals from reptiles
The first appearance of new class from mammals or birds	In the future	Elapsed time for the evolution of new class from mammals or birds is more then 220 millions

Elapsed time for the evolution of new class from mammals. In the past 220 million years there wasn't any new class of mammals. In the last 220 million years, evolving of land animals too rapidly declined. This shows that the rate of evolution of land animals has been slower over time. The cause of this declining is continental drift. If continental drift did not and all continents remain stuck together, did not reduce the rate of animal evolution, and now there was several new class, after mammals.

1. How can the gravity increase?

At the first look the increase of gravity may seem some strange, but by taking some care, we understand that it is not as impossible as it seem. The Earth can increase its gravity by several methods. In fact the Earth can increase its gravity by using for method.

One of the methods, by which the earth can increase its gravitation, is to decrease the speed of its rotation. We know that the earth rotates around itself once a day, and by the effect this rotation centrifugal force is exerted on its inhabitants. The most the speed of the rotation, the most will be the increase of the centrifugal force. And because this force is against the direction of earth's traction, when calculation the gravity of earth's surface its value is deducted from the earth's traction, and the remainder indicates the gravity on the earth's surface. So the greater the centrifugal force, the less the weight of its inhabitants, and the less this force, the more the weight of its inhabitants. If we are to accept that the gravity has increased from past up to now, we should observe the deduction of centrifugal force from the past up to now. That is we should accept that in the past the earth rotated faster around itself, and now it rotates slowly. Have such an act been done in the past really? The scientists believe that, because of the loosening of the earth's energy with the effect of friction, the rotation of the earth around itself becomes slower at the time of tides. That is the past the earth rotated faster around itself but by the passage of time its rotation became slower. So we see that such an action occurs in reality, and we can accept the increase of gravity with high degree of insurance.

Another way, by which the earth can increase its gravity, is to increase its mass. The earth is able to increase its mass by two methods: one method is to receive mass from the outside of the earth, and the other one is that this mass is added to it inside the earth. In the first method which the mass must be added from the outside, we can see the meteorite, the annual amount of which is 5 million ton. It means that this amount of material is added to the mass of the earth each year. The sun also distributes some part of its mass, in the from of particles, to the space around it, and 2.1kg of this amount is added to the earth per second. And also, the earth receives 14-ton cosmic dust from the space around. So we see that, in fact, the earth receives mass from the space surrounding it.

But what about the increase of earth's mass inside it? Assume that we are pumping a ball by a pump. In fact we increase the air inside the ball. Because the ball's shell has a certain volume, but if we want to pump it very much, the additional air will exit from its hole or it will be torn up and the air will exit. The earth's sell has a certain volume like the ball, and if the mass inside increase, the additional materials will exit through a hole or crack from the inside of the earth. Does such an action occur in reality? We know about the outflow of molten materials from the inside of the earth to its surface. These materials exit either from the volcano, which are like holes on the earth's shell, or through the long cracks, which are under the oceans. So, in fact, we observe the increase of the earth's mass. So we can accept the increase of the gravity with high degree of insurance.

The third method for the increase of the gravity is related to the following law:

"Mass of material increases by the effect of movement."

Of course this law is related to high speeds? Can the earth have high speeds? We know that there are so many galaxies in the world, and the theory of "expanding universe" says that these galaxies become far a way from each other, every moment. The speed of their going away depends on their distance. That is, the remoter the two galaxies are from each other, the higher is the speed of their going away. If the distance between two galaxies is very much, those two galaxies will go away with very high speed. If we consider tow objects in these two remote galaxies, they will be moving with very high-speed relation to each other. Now, assume that, one of these galaxies, is our Milky Way, and the other one is another certain galaxy which is very distant from Milky Way. Also assume that, one of those two objects which are on the galaxies, is our own earth which is on the Milky way, and other object is another certain sphere on the other galaxy. In this case the earth has a very high speed relative to this certain sphere, and the earth is moving with a very high speed relative to that certain sphere.

So we see that there are numerous objects in the world around us, that the earth is moving with a very high speeds relative to them. So, according to the law of the law of increase mass because of the movement with very high speed, the mass of the of the earth can increase each moment. And as result, its gravity can increase. The fourth method, by which the gravity of ears can be increased, is the decrease of the earth's ray with fixed mass. That is, the earth should be denser. In this case, the earth, like some other starts must be more compressed and smaller.

So, by approaching the surface of the earth to its center, the gravity on the surface will be more. But in fact, the evidences do not show such an action, and the earth doesn't use this method to increase its gravity.

So, we notice that, at the first look the increase of the earth's gravity seems some strange, but in fact it is possible, and this action occurs in reality, and the earth increase its gravity, gradually, by the passage of time, and we can understand the increase of the gravity by its effects on the body size of the animals.

2. The Clues for an Increasing Gravity

To be in agreement with the expanding Earth entails to be in agreement with an increasing mass of the Earth. That is, generation of elements from elementary particles and energy that reach Earth from its outer space.

I have written a rule named the rule of classification for evolution of mass in the Earth that you can find represented in the next diagram. According to this rule, the generation of substance in the Earth, from elementary particles and energy, is easily acceptable. It is like to easily accept that a molecule of water comes to existence by synthesis of hydrogen and oxygen. By accepting the rule, we can accept that substance is being generated in the Earth, causing the increase of gravity and the expansion of the Earth. The place of substance generation in the Earth is its central part, which causes a flow of magma toward the surface of the Earth. In different times – depending from the conditions of the Earth's interior – different elements come to existence (iron, copper,etc)

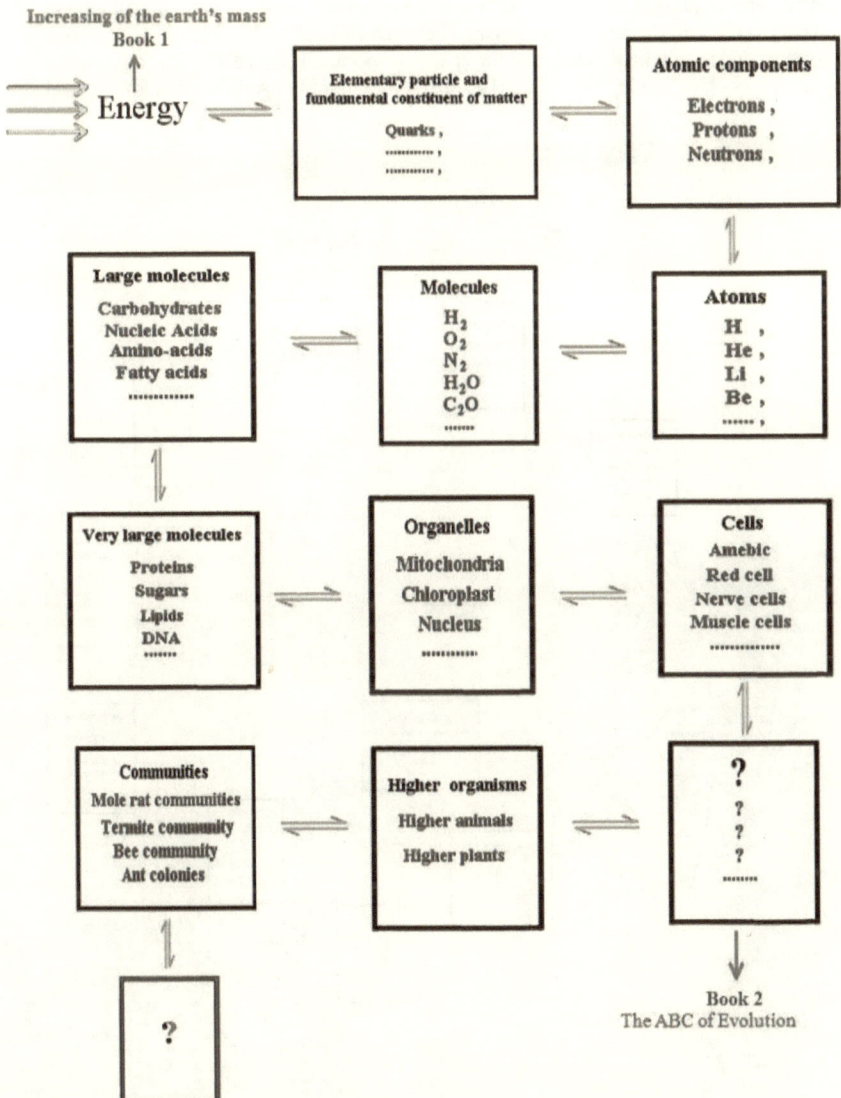

The rule of classification for evolution of mass in Earth. This figure shows the evolution of mass stage by stage, from the smallest particle to animal communities. Each step (stage) come to existence by assembling parts of the previous step. The occurrence time of each step is after the previous step. For example, proteins can only occur if Amino-acids already exist before. There are two unknown steps in the diagram – indicated by question marks. One of them is the last step that has not come to existence and will occur in the future. But one of these unknown steps is between cells and high animals/plants. This is the step I realized for the first time and scientists have not any information about that. I have mentioned many evidences for existence of this step in my second book (Mardfar, 2001).

By this rule and diagram, I could find out one of those lost chains of the path of evolution of substance in the Earth: the chain that is located between cell and high animals/plants. This is a chain that no one has thought before and never was suspected its existence. This chain is the topic of my second book (Mardfar 2001; The ABC of Evolution of high Animals/Plants)

Increase of the earth's mass

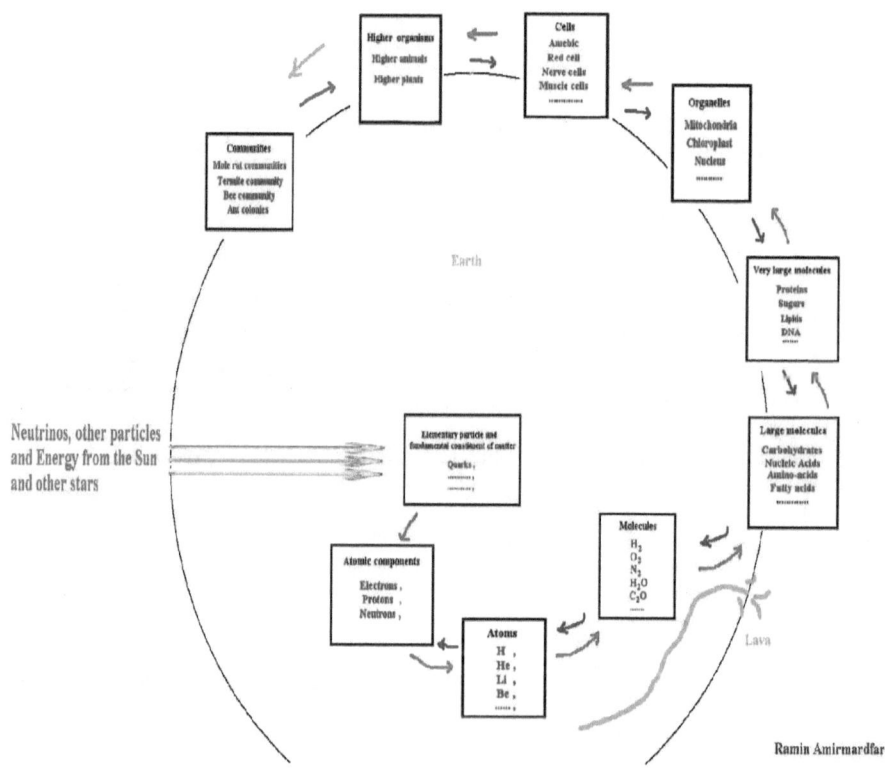

Ramin Amirmardfar

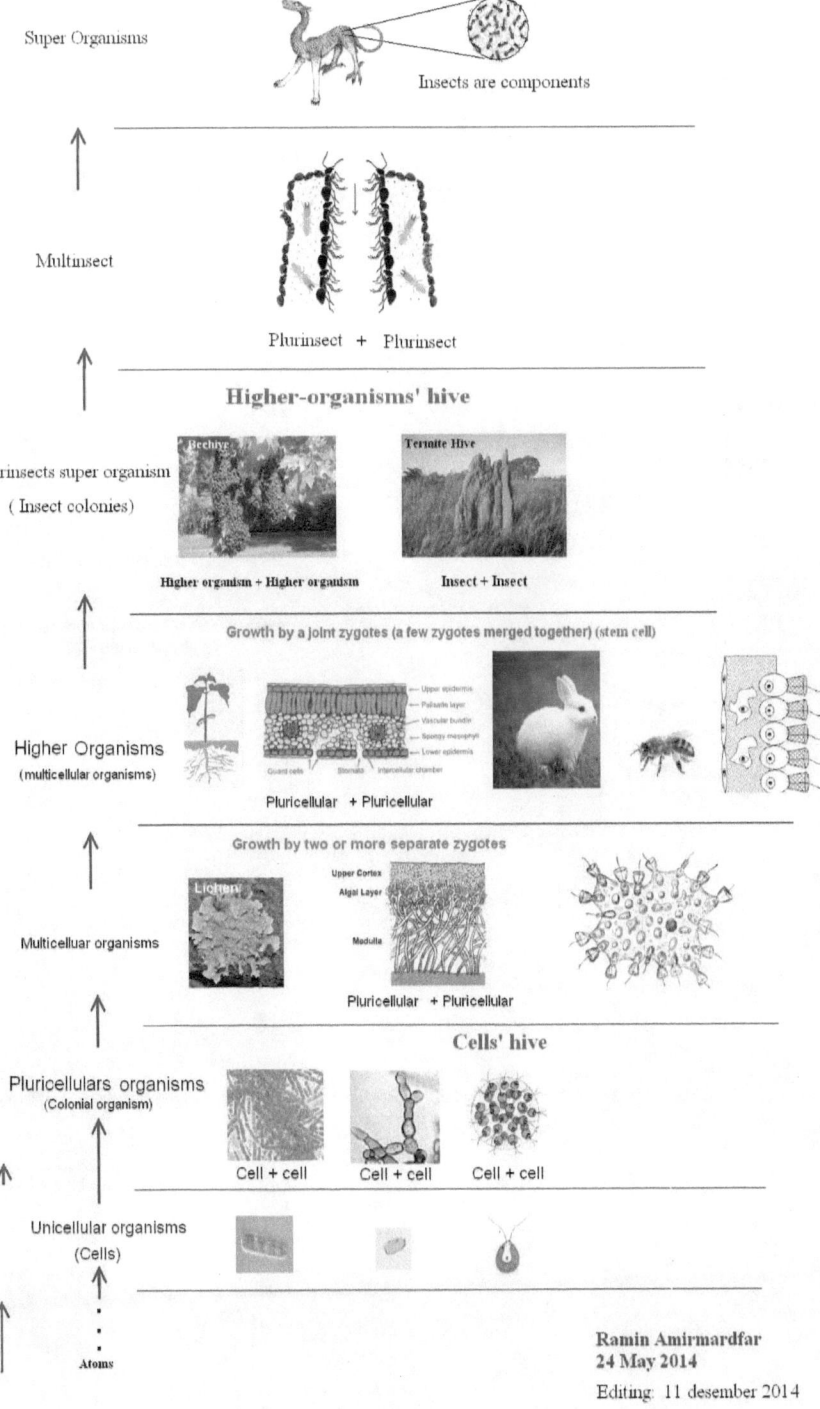

Super Organisms

Insects are components

Multinsect

Plurinsect + Plurinsect

Higher-organisms' hive

Plurinsects super organism

(Insect colonies)

Higher organism + Higher organism   Insect + Insect

Growth by a joint zygotes (a few zygotes merged together) (stem cell)

Higher Organisms

(multicellular organisms)

Pluricellular + Pluricellular

Growth by two or more separate zygotes

Multicelluar organisms

Pluricellular + Pluricellular

Cells' hive

Pluricellulars organisms

(Colonial organism)

Cell + cell   Cell + cell   Cell + cell

Unicellular organisms

(Cells)

Atoms

Ramin Amirmardfar
24 May 2014
Editing: 11 desember 2014

3. Main topics for my second book

How the appearance of the first animal stem cells?

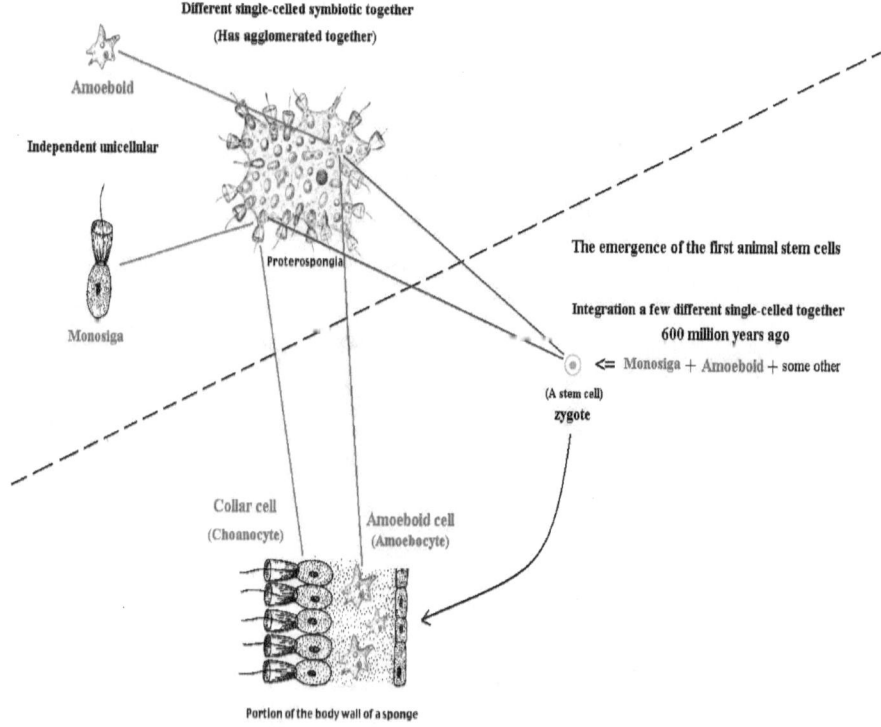

A animal "stem cell" has emerged for the first time on Earth from the merger of tow or several different independent single-celled in 600 mya.

Ramin Amirmardfar
Friday - 2016 08 July

The plant is a high-level lichen

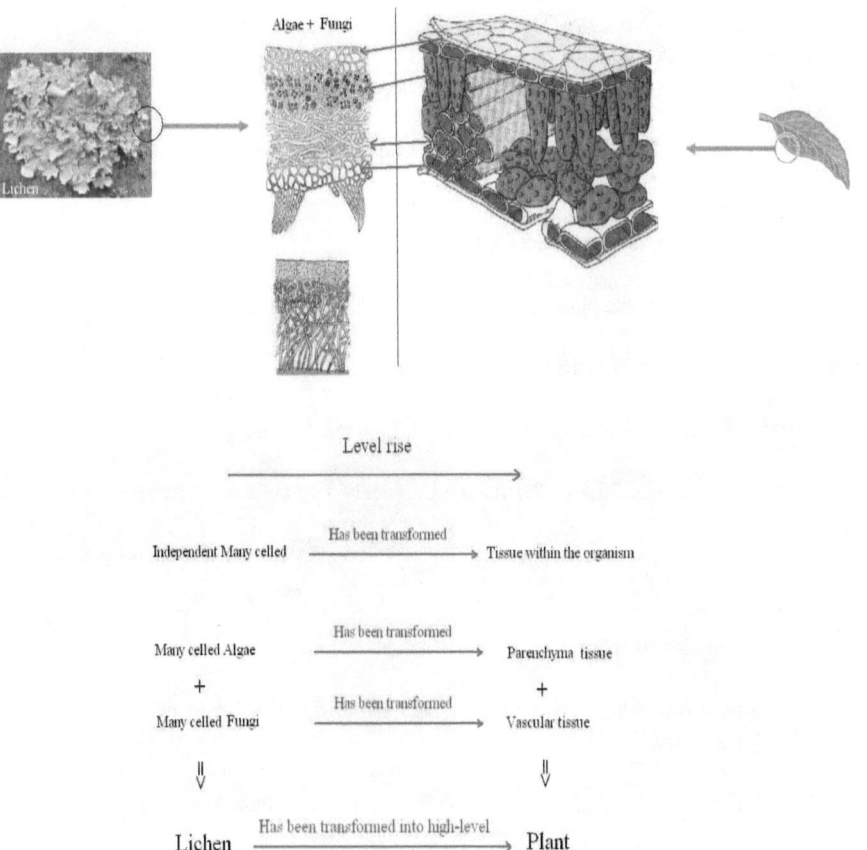

Algae + Fungi

Lichen

Level rise

Independent Many celled — Has been transformed → Tissue within the organism

Many celled Algae — Has been transformed → Parenchyma tissue

+ — Has been transformed → +

Many celled Fungi — Has been transformed → Vascular tissue

‖

Lichen — Has been transformed into high-level → Plant

Ramin Amirmardafr
12 May 2015

References

Afshar F. (1980). Paleontology. Tehran University Publications and Printing, 223 pp.

Alexopoulos C.J. (1952). Introductory Mycology. John Wiles & Sons Inc., 676 pp.

Ansari F., Abedini A. (1987) What is blood? (Published by Azar) 236 pp.

Asimov I. (1966). The Neutrino, Ghost Particle of the Atom. Published by Doubleday ICO Inc., N.Y. U.S.A, 264 pp.

Asimov I. (1972). Asimov's Guide to Science. 670 pp.

Banks H.P. (1964). Evolution and plants of the past. Cornel University, 324 pp.

Gahreman A. (1983). Basic botanical. Tehran University, 784 pp.

Gamov G. (1958). Matter, Earth and Sky. Prentice Hall, England, 662 pp.

Gardner E.J. (1972). History of Biology. Utah State University, Logan, Utah, 478 pp.

Hutchinson G.E. (1965). The ecological theater and the evolutionary play. Yale University , 139pp.

Lapo A.V. (1993). Traces of Bygone Biospheres. Synergetic Press, 356 pp.

Mardfar R.A. (2000). Relationship Between Gravity and Evolution (The Theory of the Increasing Gravity), Zeinabe Tabriz, 124 pp.

Mardfar R.A. (2001). The ABC of Evolution. Published by the Author, 58 pp.

Morteza Esmaili. (1992). Agricultural Entomology. Tehran University Publications and Printing, 581 pp.

Oparin A. (1974). The origin of Life. (New York, Plenum Press) 355 pp.

Russel B.A. (1964). The ABC of Relativity. 222 pp. Sergeev B. (London, Allen & Unwin,)

Boris sergeev. Physiology for Everyone. Moscow: Mir Publishers, (1973) 448 pp.

Shojai Mahmoud (1989). Entomology (Vol 1. Morphology and Physiology). Tehran University Publications and Printing, 430 pp.

Schmidt-Nielsen K. (1997). Animal Physiology (Adaptation and Environment). Cambridge University Press, Cambridge & New York ISBN 978-0521570985

Storer T.I., Usinger R.L., Nybakken J.W., Stebbins R.C. (1981) Elements of Zoology. McGraw-Hill, 520 pp.

Talate Habibi. (1989). Vol 1. General zoology; Vol 2. Worms & Mollusks; Vol 3. Arthropods; Vol 4. Vertebrates. Tehran University Publications and Printing, 514 pp., 413 pp., 407 pp., 651 pp.

Van der Hammen L. (1988). An introduction to comparative Arachnology. SPB Academic Publishing, The Hague, 576 pp.

Wikispaces (2012). The Mammalian Transport System. http:// bioworldcandor. wikispaces.com/ The+Mammalian+Transport+System.

Contents

www.ingramcontent.com/pod-product-compliance
Lightning Source LLC
Chambersburg PA
CBHW021014180526
45163CB00005B/1952